A Story Written *in the* Rocks

The Geology of Voyageurs National Park

Chris B. Hemstad

This book is dedicated to

Richard W. Ojakangas and David L. Southwick

Voyageurs National Park
360 Highway 11 East
International Falls, MN 56649

Published in cooperation with:

Reedy Press
PO Box 5131
St. Louis, MO 63139, USA
www.reedypress.com

Jefferson National Parks Association
One Memorial Drive, Suite 1900
St. Louis, MO 63102
www.jnpa.com

ISBN: 978-0-931056-07-9

Design by Nick Hoeing

Printed in the United States of America
13 14 15 16 17   5 4 3 2 1

CONTENTS

ACKNOWLEDGMENTS

This book is a compilation of geologic discovery that begins with the earliest geologists to the region and culminates in a continuing spirit of geologic inquiry. Although many, many geologists are responsible for deciphering the parts of the geologic story told here, this book is dedicated to Richard W. Ojakangas and David L. Southwick, who were the first geologists to conduct systematic mapping of the bedrock throughout the waterways of Voyageurs National Park. It was their original field work and research that provided the foundation for all later studies.

The staff at Voyageurs National Park would like to thank the many people whose contributions were vital to the completion of the book. Special thanks go to Chris B. Hemstad who wrote the text out of his passion and love of geology; Leland H. Grim for his many hours of helping park staff understand the geology of the park; and Tawnya Lee Schoewe for taking the information given and delivering this book in its final form.

Many talented artists and photographers provided images and illustrations, including Leland Grim; Kendell A. Dickinson; Dr. Richard Ojakangas; Dr. Robert Lillie; Raymond Klass; Allison Barnes; Thomson-Brooks/Cole; United States Geological Survey (USGS); Minnesota Geological Survey (MNGS); and National Park Service (NPS) staff.

A number of people improved this book through careful content review, editing, and locating images. They are Leland Grim, Tawnya Schoewe, and Megan Bruggeman.

Funding for the development, design, and printing of the book was generously provided by a grant from the National Park Foundation, the Geological Society of America/GeoCorps, and the Friends of Voyageurs National Park.

Finally, we thank the countless visitors who provided a constant stimulus by asking, "Have you finished the park geology book yet?"

FOREWORD

A Story Written in the Rocks: The Geology of Voyageurs National Park was written to share the story told within the rocks. This book is intended as a guide to the geologic history of the park. As it unfolds for you, numerous features will be identified in the photographs and the maps. We strongly recommend that visitors experience the landscape of the park firsthand. Seeing and touching it will help you experience and form an appreciation for the diversity of geologic features and the stories they tell.

The people of the border lakes country are linked to the landscape in one form or another. This book shows how the geology of the park has influenced and shaped the lives of people who lived here in the past, people who continue to live here, and visitors to this unique national park.

By the time you complete this book, you will be able to differentiate the various rock types present in the park. Doing this will help you visualize the Earth processes that shaped the land. You will be able to see the direction the glaciers advanced by finding their scratches and reading the granite whalebacks. You will come to understand that the sand beaches, rocky outcrops, and stones that line the park's shorelines are objects carried and deposited by the last glaciers. If you are traveling by boat, the steep-walled and narrow waterways will make you think about earthquakes and fault lines, or weak broken rock, easily carved and excavated by flowing sheets of mile-high ice. Most important, through this book, we hope you will form an appreciation of the land that will inspire you to explore Voyageurs National Park.

This book tells a story about change on a scale different from that of human life. It is a story that emerges from the depths of time, enabling you to make connections between what was and what is.

—Tawnya Schoewe

BACKGROUND

A glossary to help the reader better understand geologic terminology can be found at the back of the book. The author and contributors are as follows:

Chris B. Hemstad grew up on Rainy Lake exploring the waterways and forests of Voyageurs National Park. He earned his master of science in geology from Western Washington University in Bellingham, Washington, and his bachelor of arts in geology at Gustavus Adolphus College in St. Peter, Minnesota. He also holds an Earth and Planetary Science License from the University of St. Thomas in Minneapolis. He currently teaches science in northern Minnesota.

In 2000–2, during his time at the University of Minnesota, Hemstad codeveloped the Geologic Map of Voyageurs National Park and Vicinity, Minnesota, with Richard W. Ojakangas and David L. Southwick. To these two classic and excellent geologists he will be forever grateful, because this opportunity gave him the chance to reexplore the stunning wilderness of his boyhood and interpret the incredible story written in the rocks. Over the years, Hemstad has had ten geologic documents published on glacial and bedrock geology.

Leland H. Grim received a bachelor of arts in biology and secondary education and a master of science in zoology from North Dakota State University. He carried out postgraduate studies in geology at the University of Minnesota, Twin Cities. Lee, as he is often called, served as biology and geology instructor at Rainy River Community College from 1967 to 2000. He was also employed at Voyageurs National Park seasonally as a naturalist and biologist from 1972 to 2011.

Since then, Grim has been appointed a U.S. member of the International Rainy Lake Board of Control by the International Joint Commission (IJC) established by the Root-Bryce (Boundary Waters) Treaty of 1909; ordained a deacon of the Episcopal Church of Minnesota at Holy Trinity Episcopal Church in International Falls; and appointed a U.S. member of the International Lake of the Woods and Rainy River Watershed Task Force on Bi-National Management in the Watershed by the IJC.

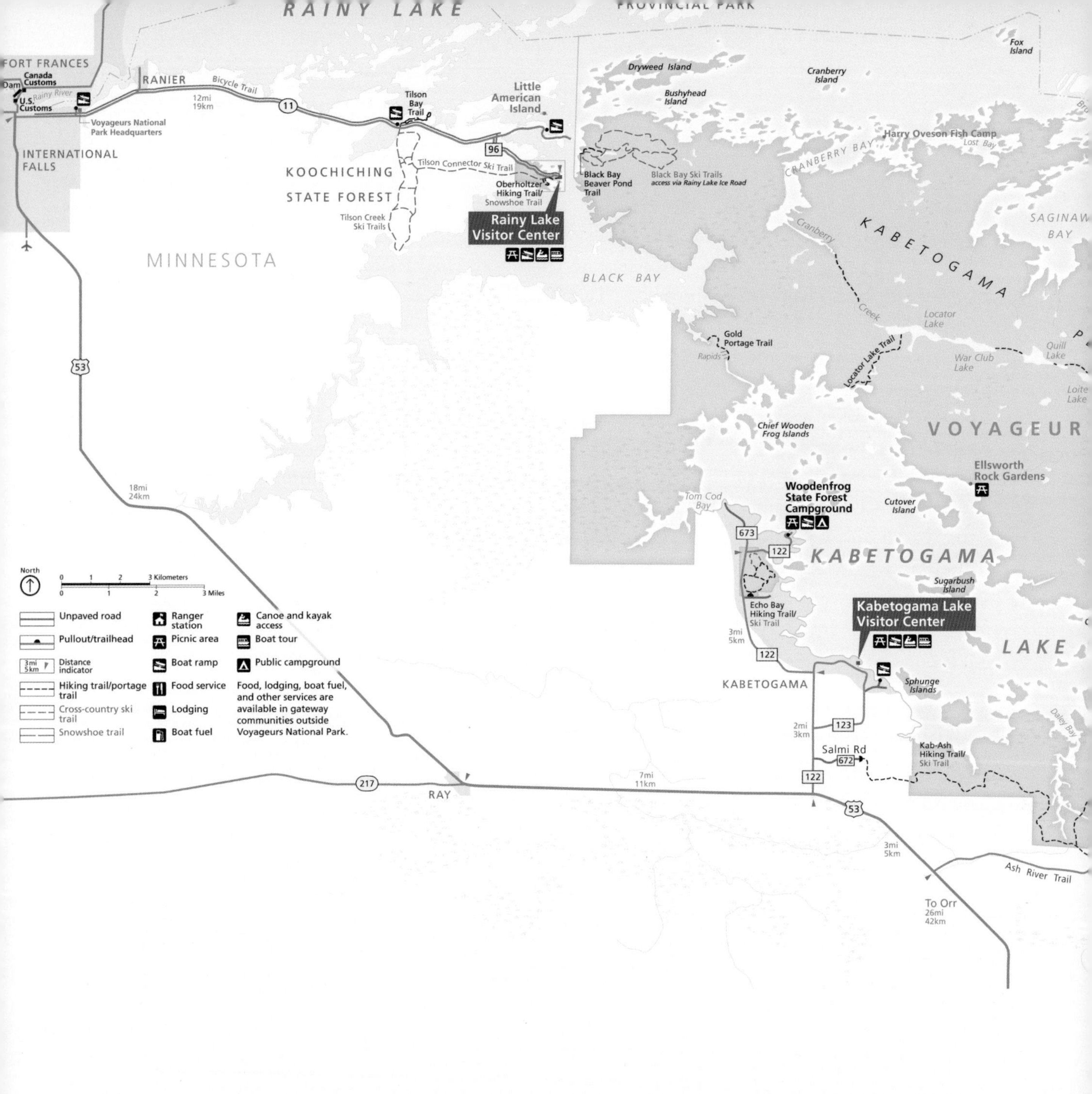
RAINY LAKE
FORT FRANCES
Canada Customs
Dam
Rainy River
U.S. Customs
RANIER
Bicycle Trail
12mi
19km
11
Voyageurs National Park Headquarters
INTERNATIONAL FALLS
Tilson Bay Trail
Little American Island
96
Tilson Connector Ski Trail
KOOCHICHING
STATE FOREST
Oberholtzer Hiking Trail/
Snowshoe Trail
Tilson Creek Ski Trails
Rainy Lake Visitor Center
MINNESOTA
Dryweed Island
Bushyhead Island
Cranberry Island
Fox Island
Harry Oveson Fish Camp
Lost Bay
CRANBERRY BAY
Black Bay Beaver Pond Trail
Black Bay Ski Trails
access via Rainy Lake Ice Road
Cranberry
Creek
KABETOGAMA
SAGINAW BAY
BLACK BAY
Gold Portage Trail
Rapids
Locator Lake Trail
Locator Lake
War Club Lake
Quill Lake
Loite Lake
Chief Wooden Frog Islands
VOYAGEUR
Ellsworth Rock Gardens
Woodenfrog State Forest Campground
Tom Cod Bay
Cutover Island
673
122
KABETOGAMA
Sugarbush Island
Echo Bay Hiking Trail/
Ski Trail
Kabetogama Lake Visitor Center
LAKE
3mi
5km
KABETOGAMA
Sphunge Islands
2mi
3km
123
Salmi Rd
672
Kab-Ash Hiking Trail/
Ski Trail
Daley Bay
53
18mi
24km
217
RAY
7mi
11km
3mi
5km
Ash River Trail
To Orr
26mi
42km
North
0 1 2 3 Kilometers
0 1 2 3 Miles
Unpaved road
Pullout/trailhead
Distance indicator
Hiking trail/portage trail
Cross-country ski trail
Snowshoe trail
Ranger station
Picnic area
Boat ramp
Food service
Lodging
Boat fuel
Canoe and kayak access
Boat tour
Public campground
Food, lodging, boat fuel, and other services are available in gateway communities outside Voyageurs National Park.

Moose Bay
RAINY LAKE
CANADA
UNITED STATES
Big Island
Blueberry Island
Kempton Channel
Browns Bay
Camp Marston
Finger Bay
Anderson Bay Overlook Trail
Peary Lake
Anderson Bay
Ryan Lake
ATIONAL PARK
Brown Lake
Beast Lake
Beast Lake Trail
MICA BAY
Kettle Falls Historic District
Hotel
American Channel
Canadian Channel
Mica I
Kettle Falls Dam
Portage via vehicle
Squirrel Falls Dam
Little Shoepack Lake
Jorgens Lake
Cruiser Lake
Cruiser Lake Trail
Quarter Line Lake
Ek Lake
Voyageurs Narrows
Agnes Lake
LOST BAY
Kubel Island
Tar Point
NAMAKAN LAKE
Namakan River
Ash River Visitor Center
Blind Ash Bay Hiking Trail/ Snowshoe Trail
Blind Indian Narrows
I.W. Stevens Resort
Fox Island
Blind Pig Channel
Your Island
Pike Island
Namakan Narrows
Kabetogama Lake Overlook
Sullivan Bay Snowshoe Trail
Old Dutch Bay
Beaver Pond Overlook
Kab-Ash Ski Trail
Hoist Bay
Junction Bay
Moose River
3mi 5km
Voyageurs Forest Overlook
Kab-Ash Hiking Trail
ASH RIVER
Ash River State Forest Campground
1mi 2km
129
Ash River
Ash River Falls
KABETOGAMA STATE FOREST
Burnt Island
SAND POINT LAKE
Little Trout Lake
Grassy Bay Cliffs
GRASSY BAY
Browns Bay
Harrison Narrows
Canada Customs
PORTAGE BAY
Mukooda Lake
Northwest Bay
Casareto Cabin
King Williams Narrows

A Story Written *in the* Rocks

Figure 1. Volcanic eruption, Hawaii Volcanoes National Park.

Chapter 1

INTRODUCTION

> *The shatter'd Rocks and Strata seem to say,*
> *"Nature is old, and tends to her Decay":*
> *Yet, lovely in Decay and green in Age,*
> *Her Beauty lasts to her latest Stage.*
> —Mary Chandler, "The Description of Bath"

Imagine you are sprawled on a bare rock that is on the edge of a looming, primordial ocean. Suddenly, the top of a massive conical-shaped volcano explodes, blowing itself into the water below. Your stomach churns as you feel a massive earthquake shake the ground and you see a gushing column of red-hot ash and debris rocket skyward. It is infused with an ethereal blue glow produced by lighting of its own making. The top of the blazing column quickly begins to form a cascading mushroom cloud. Lit by flickering lightning, it is large enough to rival any ever produced by nuclear explosions. Seconds later, an eardrum-snapping detonation rolls over you, bathing the landscape in the primordial roar produced by the escape of super-pressurized gases and red-hot, liquid rock. You are witnessing the first stage in the story of how this area now known as Voyageurs National Park was created (fig. 1).

How did the landscape of this national park come to be? What were the forces of nature that contributed to the shaping of it? How old are the rocks that formed the rugged landscape upon which everything depends? And how do these natural processes affect us today? People have been asking these questions for generations. The answers deeply define who we are as humans and how we interact with the landscape. The answers surround us and we can unveil them if we look carefully. Look down at the Earth you are walking on. What do you see? What color is it? Touch it: Is it smooth or rough? Do the rocks remind you of other places that you have visited before? If you compile answers to these questions, you begin to tell the story of this park. With further viewing and investigation, you will soon find yourself immersed in an ever-broadening and intriguing story.

The geologic story of Voyageurs National Park relates how earthquakes and mountains, water and ice, liquid rock and jack pine roots, have shaped the landscape, provided the foundation for the present ecosystem, and influenced human cultures. This is a story that begins long before humans, indeed even before plants and animals existed. You might then ask how we know this. As you read on you will discover that the story is written in the rocks.

The rocks of the park preserve some of the oldest records of the Earth's history. By observing and learning about these rocks you can learn about how the early continents were formed. The rocks tell you about an aged world dominated by volcanism, earthquakes, and rapid mountain building. The more you read and the more you explore, the more your vision into the past expands. Eventually, time speeds up, and millions of years roll by like boxcars on a passing train.

This book not only shares the story of the park's rocks but presents a vastly different landscape from the one we know today.

The accumulated knowledge and imagination of people who have been listening to and retelling the story for generations play a key role. How you make sense of this landscape depends upon your imagination. Throughout the past hundred years, many geologists, with specialized tools and techniques, have discovered new chapters within the rocks of the park. Using this information, they have been able to combine imagination and scientific reasoning to reconstruct an ancient landscape—one of primitive oceans, steaming volcanic islands, and rumbling earthquakes—interwoven within a fascinating tapestry of colliding continents.

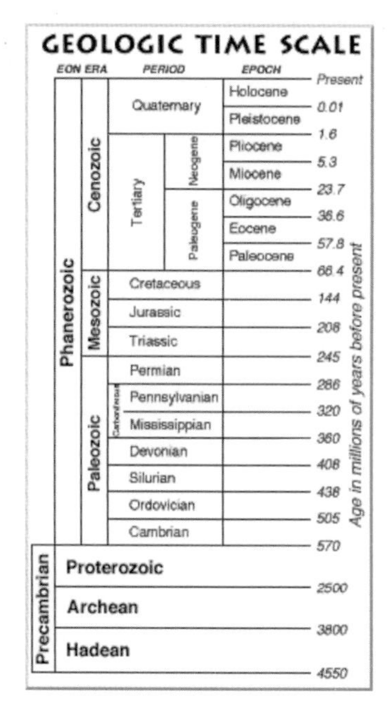

Figure 2. Geologic Time Scale, USGS.

Interestingly enough, every rock seems to tell a particular story. Some of the stories do not match with what other rocks have told us. Are the rocks telling lies? No. Geologists sometimes misinterpret what they find because there is not enough evidence for a complete story. How then, can you make sense of this landscape and the story written in the rocks? In both reading and learning, imagination is the key.

Like any good story, as time goes on, this one changes and adapts as more and more evidence mounts and geologic theories become more sophisticated. As our exploration of the rocks continues, it delves further into the past. What begins to emerge is an ever-changing landscape shaped by continual interactions and processes.

The park's story begins 2.7 billion years ago, during the Archean Eon of the Precambrian Era (fig. 2). Earth itself is about 4.6 billion years old. We will consider in detail only two time periods, the Archean (3.6–2.1 billion years ago) and the Pleistocene/

Holocene (2 million years ago to present). During the in-between years there were no major geologic events occurring other than erosion that “sandpapered” the high mountains to a lower flat plain.

The geology of the park has shaped cultures and influenced the movement of people over the landscape. For thousands of years the waterways and lakes of the region have provided a reliable route of travel through an otherwise virtually impassable wilderness. Gold was found in the rocks of what is now known as Voyageurs National Park, initiating a gold rush that brought thousands of people to the border lakes country. After the failure of the gold mines, logging became the major industry and continues to provide the economic framework of the region today. Although the lure of gold and large pine still lingers, it is the beauty and history of the area that finally led to the designation of Voyageurs National Park.

Chapter 2

THE HEART OF THE CONTINENT

> *The result, therefore, of our present enquiry is, that we find no vestige of a beginning, no prospect of an end.*
> —James Hutton, *Theory of the Earth*

Voyageurs is the only national park where a visitor can see rocks that tell the story at the heart of the North American continent. This landscape is a significant part of our nation's heritage. In fact, the legislation that authorized the founding of Voyageurs National Park specifically states the goal of the park as being "to preserve, for the inspiration and enjoyment of present and future generations, the outstanding scenery, geological condition and waterway system which constituted a part of the historic route of the voyageurs who contributed significantly to the opening of the Northwestern United States." To encourage a full appreciation of what this statement means, let us explore it in more detail.

4.6 Billion Years Ago (BYA). Throughout the margins of this book, the Voyageurs' Geologic Clock tracks our "geologic day," which scales geologic time to a twenty-four-hour time period.

The geologic story of the park lies within the Earth's history as shown in figure 2. Evidence suggests that the Earth formed 4.6 billion years ago. Prior to about 3.8 billion years ago, there is very little evidence about what was happening on Earth. If there were significant landmasses prior to this time, evidence of them has disappeared. Starting about three billion years ago, however, the

Anderson Bay, Rainy Lake, NPS.

continent we call North America began to come into existence. What makes the park's geology significant is that the rocks date from a time when the crust of the North American continent began to rapidly form and evolve. The oldest rocks within the park are at least 2.7 billion years old.

Before exploring the formation of the continent, let us orient ourselves in geologic time. The park's rocks were formed in a time almost too ancient for the imagination to grasp, when oxygen was nearly absent in the atmosphere and the internal temperature of the Earth was much hotter. Heat flow rates are believed to have been approximately three times higher than today. In addition, the Earth was spinning faster, producing days of about fifteen hours in duration.

We are used to thinking of time in the context of a human life—in minutes, hours, days, even hundreds of years. Imagine all you have experienced in the past year, all the joys and frustrations, all your ups and downs, all the 365 days of your morning rituals: waking, showering, getting dressed, and leaving the house.

To make the switch from a human life span to geologic time, we must stop counting in years and begin to count in hundreds of thousands, millions, even billions of years. From a human perspective these lengths of time are hard to comprehend. For example, a million years is one thousand periods of one thousand years each and a billion years is one thousand periods of one million years. Geologic time soon becomes overwhelming.

A common analogy is to scale all of geologic time to our normal twenty-four-hour day. That is, imagine that the entire history of the Earth is one day. We divide our days into smaller periods of time—hours and minutes—and to these smaller periods we occasionally ascribe names, such as morning, noon, afternoon, early evening, and so on. Geologists have likewise split geologic time into smaller divisions—eons, eras, periods, and epochs. These divisions mark significant events or changes in the history of the Earth.

According to our "geologic day" analogy, most of Earth's history, about the first twenty-one hours (midnight–9:00 p.m.), takes place during a period of time called the Precambrian (fig. 2). Evidence suggests single-celled life came into existence in the first six hours of our "geologic day" (3.5 billion years ago). In our "geologic day," the events that followed were the explosion of multicellular life, and they took place during the final three hours of our fictitious day.

Looking at figure 2, you will no doubt notice that the time following the Precambrian is divided into many more segments than the Precambrian itself. This seems unreasonable, given that the Precambrian period occupied most of the world's history. However, we can only divide and classify periods of time for which we have evidence.

Now, let us place the rocks in the park and ourselves in a one-day (twenty-four-hour) clock. At 8:20 a.m., the oldest rocks within the park, pillow lavas, were spilled and formed on an ancient seafloor. The North American continent grew together between 10:00 and 11:00 a.m. At 1:30 p.m. the youngest rocks in the park—dikes of gabbro, basalt, and andesite—shot through older rocks. Forty seconds before midnight are hominids. One-tenth of one second before the twenty-fourth hour we find writing in the cultures of North Africa and the Middle East.

Geologic time is so impressive because during its great expanses of millions of years, events that seem impossible can occur. The continents can skate hundreds of miles across the Earth's crust, mountains can rise thousands of meters in height, and continents can be sutured together from separate parts like handfuls of clay. In the depths of geologic time, all that we think of as permanent—continents, mountains, and rocks—is in actuality plastic, easily created, bent, and destroyed.

Chapter 3

THE CANADIAN SHIELD

Now that we have oriented ourselves in geologic time, let us get our bearings in geologic space. The park lies atop an area of exposed Precambrian rocks called the Canadian Shield (fig. 3a). Surrounding the Canadian Shield to the west and south are areas of Precambrian rock that have been covered by sediment that is called a platform. Taken together, the Canadian Shield and the platform are called a craton. A craton is an area of the world's surface that has been geologically inactive for a long period of time: no rising mountains, no volcanoes, and no high-intensity earthquakes. Cratons form the core of each of the continents.

What does the Canadian Shield look like? The back cover photo shows the Canadian Shield in the park—a rocky landscape that has been worn to its present elevation by billions of years of erosion. The advance and retreat of glaciers during the past two million years scraped away any trace of ancient soil and scoured shallow holes that are now filled by lakes, ponds, and rivers: hence Rainy Lake, Kabetogama Lake, Namakan Lake, Sand Point Lake, and twenty-six interior lakes (see park map on pages x-xi). Since the last glaciers melted, thin soils formed, and flora and fauna

Figure 3a. North American Craton. Voyageurs National Park lies on the southern edge of the continental shield, a vast region of very old igneous and metamorphic rocks. To the south, continental shield rocks are covered by younger sedimentary layers of the continental platform. Relatively young mountain ranges surround the craton. Modified from Stephen Marshak, *Earth: Portrait of a Planet* 2d ed (W. W. Norton and Company, 2005).

recolonized the barren landscape. The area that includes the park is now covered in a transition zone where hardwoods and conifers meet and mix, otherwise known as the northern boreal forest.

If we begin to examine the details of the Canadian Shield, the first thing we notice is that it is a series of continuous bands of parallel rock (belts) running roughly east to west (fig. 3b). Each belt contains rock of a similar age and origin and is different from the belt to the north or south of it. In the region of the park, we have, from north to south, the English River, Wabigoon, Quetico, and Wawa belts. You may notice (fig. 3b) there are different colors to illustrate the belts. The different color belts represent different types of rocks. Greenstone belts contain rocks formed from the cooling or hardening of molten lava, subsequently changed by metamorphism (altered under heat and pressure). The Wabigoon and Wawa belts are greenstone belts. A second type of belt, called a metasedimentary belt, usually separates greenstone belts from each other. The word "metasedimentary" comes from "metamorphosed" plus "sedimentary." Metasedimentary belts consist of rocks that were initially formed from sedimentary materials such as sand, silt, clay, pebbles, and cobbles and include siltstone, sandstone, conglomerate, and shale, among others. The sedimentary rocks have been metamorphosed to rocks such as schist and quartzite. The Quetico and English River belts are examples of metasedimentary belts.

These alternating greenstone and metasedimentary belts are bounded on the north and south by older areas of gneiss (rock formed by the metamorphism of granite; fig. 3c). On the map, these older areas are labeled as the Minnesota River Valley and the Pikwitonei subprovinces. These older rocks existed before the park's story began and geologists do not know for certain the origins of these older rocks. As we will see later, these older landmasses were cemented to younger landmasses to form the continent's core.

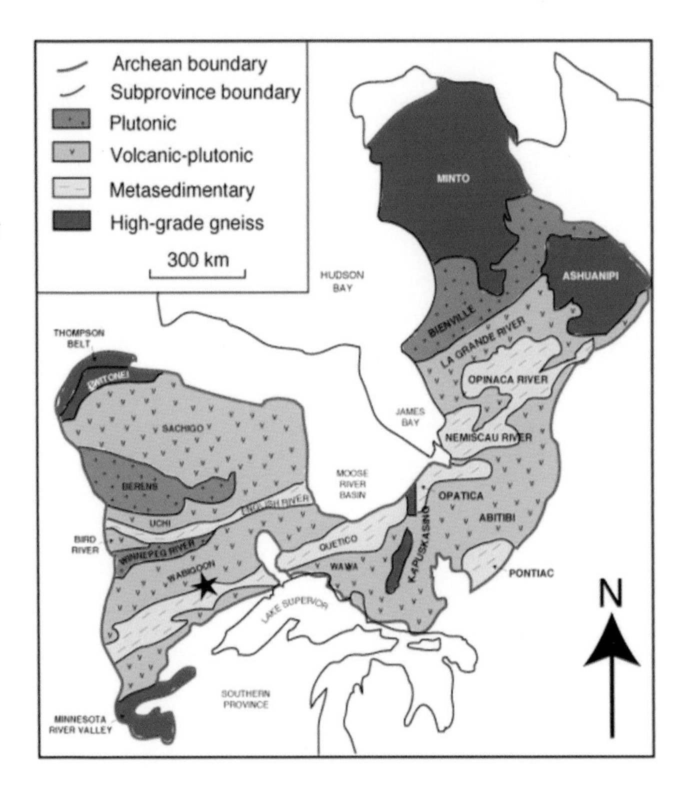

Figure 3b. Map showing the location of the Superior Province of the Canadian Shield. The star represents the location of Voyageurs National Park along the contact between the Wabigoon and Quetico belts of the Superior Province, USGS.

All branches of science use specialized terms to describe objects or phenomena, although it may seem to the reader that these terms were developed to serve as a dense and thorny thicket. The truth is that many of the terms were chosen because scientists considered them descriptive and accurate. In this book we will keep the use of specialized terms to a minimum. However, we will introduce a few that may help the reader approach and understand geologic literature. For clarification on any of the terms mentioned in this book, reference the glossary.

Figure 3c. Gneiss in a leucogranite outcrop at the Cruiser Lake Trailhead near Lost Bay, Kabetogama Lake, NPS.

The first group of terms are those geologists use to divide the landscape. Geologists have divided the Canadian Shield into provinces because each province has a unique geologic story. The Superior Province underlies the park (fig. 3a) and extends from Manitoba to Labrador. The Superior Province is, in turn, divided into subprovinces. Each subprovince is composed of rocks of a similar age and origin. Each of the greenstone and metasedimentary belts we previously described are their own subprovince. Thus, we have the Quetico and Wabigoon subprovinces in the park.

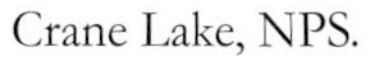

Crane Lake, NPS.

The characteristics of the Canadian Shield (or Superior Province) and its craton are not restricted to North America. A look at a geologic map of the world shows that each continent has one or more shields and cratons. Not only that, research has shown that these shields are generally made up of an alternating series of greenstone and metasedimentary belts. Most of the world's shields date to between 3 and 2.5 billion years ago. Citing this and other evidence, geologists suggest that this period was when the crust of our existing continents was rapidly forming. Though there are various theories to explain how the continents formed and why they are composed of alternating belts, the one that has gained the widest acceptance is plate tectonics.

Chapter 4

THE THEORY OF PLATE TECTONICS

According to the theory of plate tectonics, the Earth's crust is partitioned into numerous plates that "float" on the Earth's semi-molten mantle (fig. 4).

These plates can be composed of continental crust, oceanic crust (otherwise known as a lithosphere), or a combination of both. Continental crust consists of the rocks that make up the continents and averages 100 kilometers thick; oceanic crust is composed of the rocks underlying the seafloor. Geologists believe that currents of heat, called thermal convection currents, deep within the Earth, put the plates in motion. To understand a thermal convection current, take a look at the pot of water next time you boil an egg. You will notice that heat from the burner sets up swirling currents within the pot. At the point where currents meet one another (usually in the middle of the pot), bubbles and waves of water are pushed up above the surface. Convection currents within the Earth work on a similar principle, although the process is much slower. Where convection currents meet, places called mid-ocean ridges, heat is concentrated and

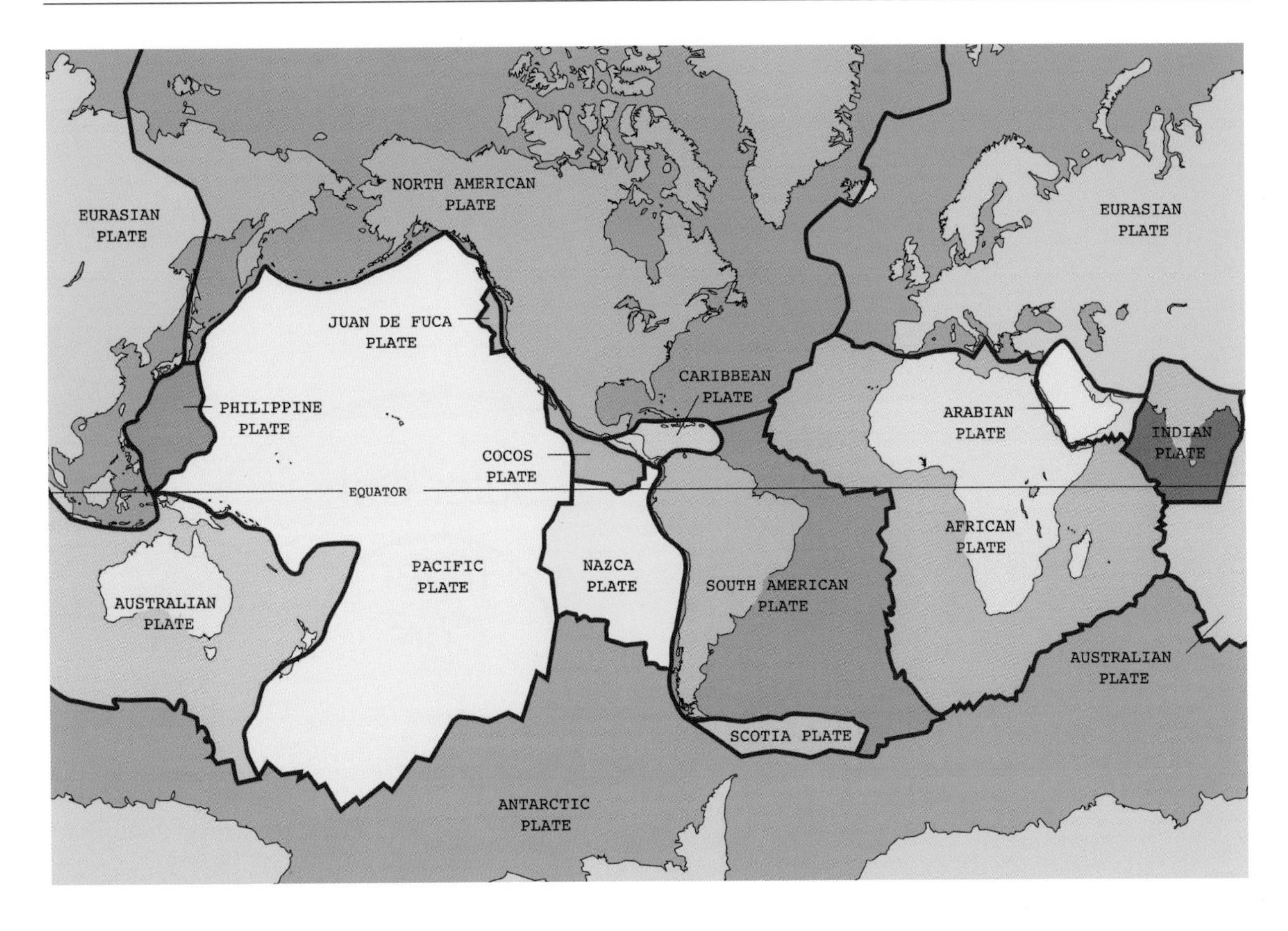

Figure 4. World map showing major tectonic plates, USGS.

molten rock wells upward. The molten rock, or magma, pushes the oceanic crust apart, creating a spreading zone, an expanding or rifting area on the oceanic plate, also known as a mid-oceanic ridge (fig. 5).

The surface of the Earth is a limited space: One might surmise that if the crust is expanding in one part of the world, it must be

Figure 5. Convection currents pushing open a mid-oceanic ridge, USGS.

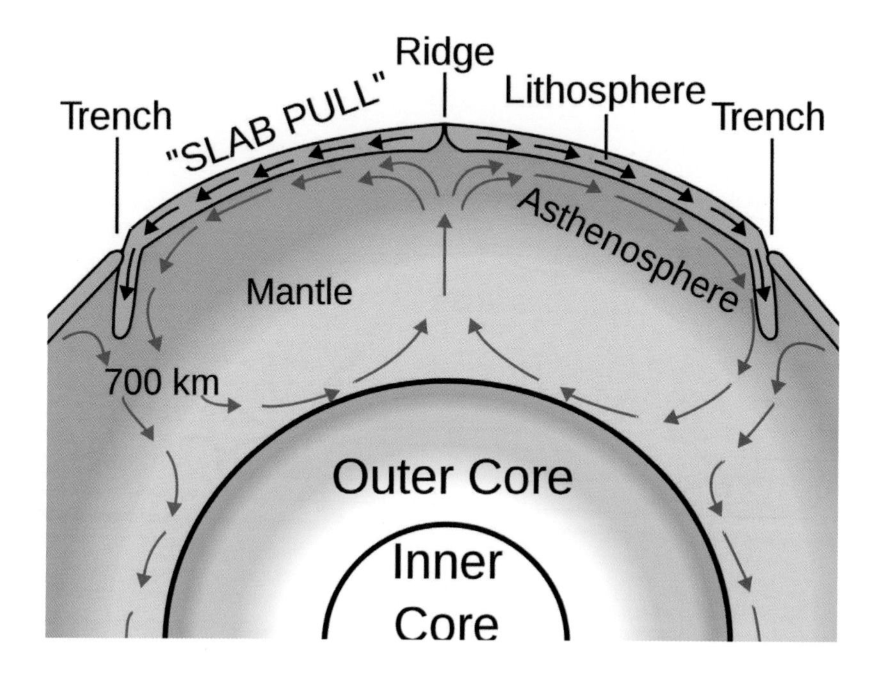

colliding and contracting at another location. Indeed, this is what appears to be happening. Many of the Earth's plates are colliding with one another. These collisions are very important because a number of the geologic processes that shape our world take place in these zones of collision. Volcanoes, earthquakes, mountain building, the deformation of continental plates, and metamorphism commonly result from colliding plates.

The geologic events that occur during plate collision depend upon the relative density of the plates at their point of contact. Oceanic crust is largely made of a very dense volcanic rock called basalt. Continental crust contains a host of rocks that are less dense. As a consequence, oceanic crust is heavier than continental crust. If oceanic and continental crusts collide, the oceanic crust

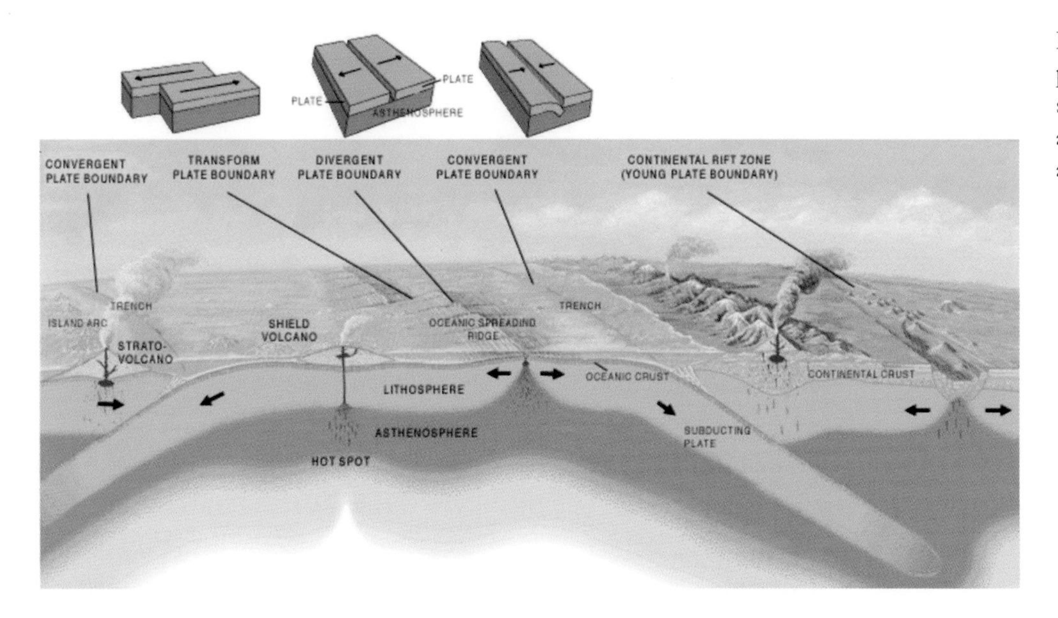

Figure 6. Colliding plates creating a subduction zone and volcanic activity, USGS.

will sink beneath the continental crust, creating what is called a subduction zone (fig. 6). If two oceanic crusts collide, the heavier oceanic crust will subduct. However, if continental crusts collide, neither plate sinks.

The temperature of the Earth generally increases from the surface toward the core. As subducted, or sinking, oceanic crust plunges deeper into the Earth, it is heated and may eventually melt. Ocean water that has been trapped and carried downward with the crust is thought to aid this process. When heated, this water becomes water vapor, which, under pressure, helps melt rock. Melted rock, or magma, is less dense than solid rock. As a consequence, molten rock will migrate or "float" upward through zones of weakness in the rocks above it, generally collecting in magma chambers at various depths below the surface. If the pressure is sufficient to rupture overlying rocks, molten rock or magma will spill, as lava, onto the Earth's surface—and a volcano will be born.

Subduction zone boundaries experience numerous earthquakes. Great stresses build up during the collision of plates. Small movements and readjustments of the Earth's crust (earthquakes) eventually relieve these stresses. The movements are small on a global scale but catastrophically powerful on a human scale.

Subduction also causes the folding and faulting of rocks. During subduction, marine sediments are scraped off the subducting, or sinking, oceanic plate. The forces of collision can break up these sediments into a jumble that looks similar to stacked dominos. This is called a mélange. Some of these sediments may also be carried downward by the subducting plate and metamorphose under heat and pressure into metasedimentary rocks.

Continental crust is not dense enough to subduct below another piece of continental crust. It is too light and buoyant. When continental crusts collide, the collision generally creates mountains. To imagine how this may happen, think of a square layered chocolate and vanilla cake on the table in front of you. Cut the cake in two, but do not separate the halves. With your left palm flat against the left side and your right palm flat along the right side, begin pushing your hands together. The center of the cake and everything between your hands begins squeezing together. The horizontal layers start to lift and fold. With continued pressure, they finally crumple and break under the perpendicular compressional force you are applying. The center of the cake will begin to uplift to form "chocolate and vanilla mountains" between your palms. Geologists call a mountain building event an orogeny. You have produced a "chocolate and vanilla orogeny" between your hands!

Another process likely to result from the collision of any two blocks of crust is metamorphism. When heated or squeezed, rocks become plastic. At first, they do not melt, but the atoms in the original minerals may rearrange themselves to form new minerals.

Grassy Bay Cliffs, Sand Point Lake, NPS.

Colliding plates present two possible mechanisms for metamorphism. During subduction, the sinking plate can carry sediments thousands of meters under the Earth's surface. At these depths, temperatures are higher. The weight of thousands of meters of overlying rock can create immense pressure. Under these conditions, rocks are likely to metamorphose.

The second mechanism is related to the magma chambers described above. When magma rises into surrounding rock, it is hot enough to metamorphose the surrounding rock. We call this kind of metamorphism "contact metamorphism."

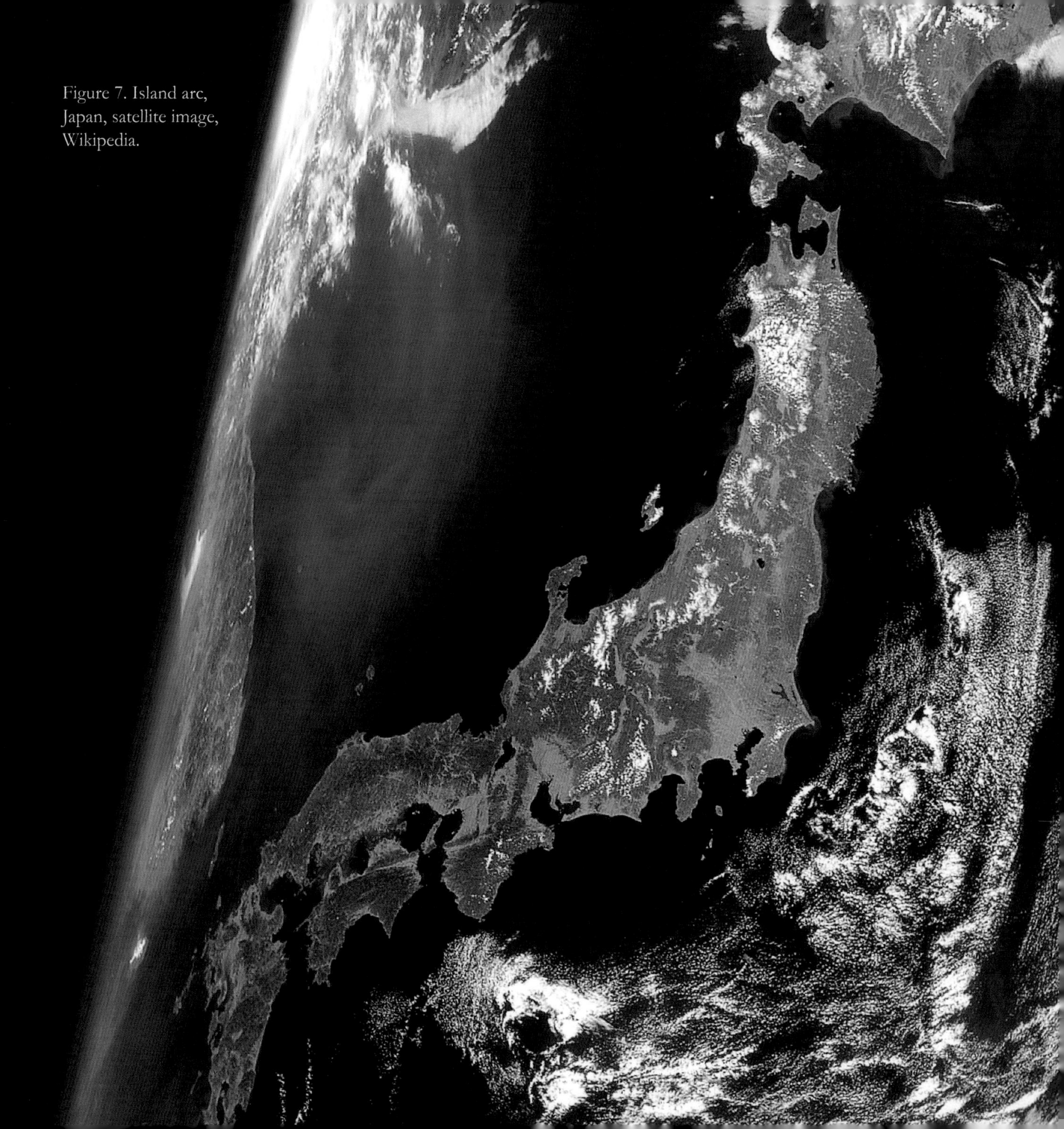

Figure 7. Island arc, Japan, satellite image, Wikipedia.

Chapter 5

PLATE TECTONICS AND THE FORMATION OF THE SUPERIOR PROVINCE

What do tectonics and the geologic processes associated with plate tectonics have to do with the geology of Voyageurs National Park or the formation of the North American continent? Evidence suggests that a series of plates collided with one another, one after the other like train cars clanging together when the locomotive stops, to form the core of the North American continent. Each time two plates collided, volcanoes were born. Earthquakes shook the Earth's surface. Sediments were folded, faulted, and metamorphosed. Molten rock formed magma chambers beneath the surface of the Earth, and mountains rose.

The plates of 2.7 billion years ago did not look like those portrayed in figure 4. While geologists differ in their opinions

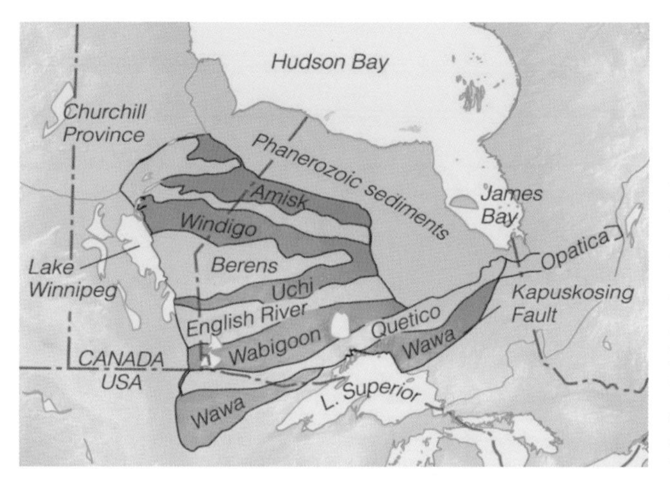

Figure 8. Map of volcanic (dark green) and sedimentary (light green) belts of the Superior Province, Thomson-Brooks/Cole © 2004.

about the exact properties of plates dating from Precambrian time, they generally agree that plates were much smaller and their movements were much faster than plates today. The plates were smaller because not enough time had passed to create large blocks of continental crust. Many of these small plates probably carried not continents, but linear chains of volcanically active islands called volcanic island arcs, similar to today's Philippine, Japanese, or Indonesian island chains (fig. 7). The plates moved faster, even though it had been cooling over the past 1.9 billion years since its creation. A hotter Earth fueled more powerful and faster convection currents deep within the Earth, thus faster movements.

The evidence to support this theory of the formation of the North American continent comes from the ages of the belts

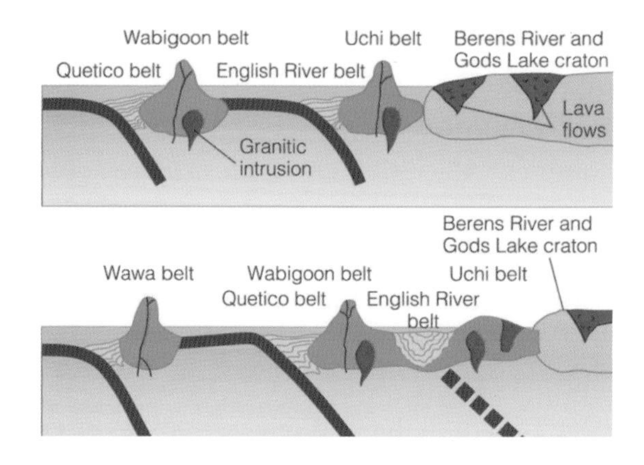

Figure 9a and 9b. Cross sections showing sequential collision of island arcs and intervening ocean sediments to form the Superior Province. Voyageurs National Park is located along the contact between the Wabigoon Belt and the Quetico Belt. The Quetico Belt represents submarine ocean sediments squeezed between the Wabigoon and Wawa volcanic island arcs during collision of the Wawa Belt with the southern boundary of the Superior Province, Thomson-Brooks/Cole © 2004.

or subprovinces that make up the Superior Province (fig. 8). Geochronologists found that the greenstone and metasedimentary belts of the Superior Province range in age from about 3.1 billion years to 2.6 billion years. The age of individual belts generally decreases from north to south. This suggests that northern belts first collided with the older Pikwitonei gneiss belt and were followed by the later collision of southern belts.

Ken Card of the Geological Survey of Canada has proposed a chronological sequence for the collision of belts within the Superior Province. He suggests that approximately 2.7 billion years ago, the Wabigoon Belt, which had already collided with the Winnipeg and English River belts, slammed into the Uchi-Sachigo Belt (fig. 9a). About 10 million years later, around 2.7 billion years ago, the Wawa/Vermilion volcanic island arcs collided with the southern margin of the Wabigoon terrane (fig. 9b). The park lies directly between the Wabigoon and Wawa/Vermilion belts and would have borne the brunt of the collisions and consequently, its rocks were heavily influenced by these events (fig. 9c).

Figure 9c. Folded sediments, Dryweed Island, Rainy Lake, NPS.

Chapter 6

PLATE TECTONICS AND THE FORMATION OF VOYAGEURS NATIONAL PARK

The next step to understanding the park's rocks is to consider the geology of each of the plates that collided to form the Superior Province. Due east of Little American Island, just to the south of Steamboat Island (see park map insert on pages x-xi), there is a small oblong island. In the exposed outcrops along the shoreline are subtle, bulbous shapes, which look like squished and contorted watermelons. Look closer and even a faint rind is visible (fig. 10). These are called pillow structures, and they are made of a type of volcanic rock called basalt. These pillow shapes tell us that the basalt cooled under water. Pillows are formed either by the bubbling of blobs of lava underwater or by branching flows advancing in stepwise fashion. Pillow

Figure 10. Pillow structures on Rainy Lake, Ontario, Canada, © Leland H. Grim. While these are larger in size than the ones found within the park, they are similar in shape to those seen on a small oblong island south of Steamboat Island, Rainy Lake.

lavas usually form at mid-ocean ridges where intruding magma is pushing two oceanic plates apart and erupting as lava on the ocean floor.

Pillow basalts formed anywhere from 3.1 to 2.7 billion years ago and are the oldest rocks in the park. We know they fall in the range of those years because of their location below rocks shown to be 2.7 billion years old. These pillows probably had more of an oval structure when they were formed. The force of colliding plates squished and stretched these pillows into oblong shapes with their long axes oriented east and west.

The calm submarine eruptions of lava and the formation of pillow basalts constituted one of the first steps in the evolution of the plates that eventually formed the Superior Province and the park's bedrock. Geologists believe the eruption of pillow lavas probably piled layers of lava until it became very thick.

Because the Earth is a finite space, tectonic plates can only rift and expand so much before they begin to collide with adjacent plates. Geologists believe that rifting eventually pushed adjacent plates into collision. During the beginning process of subduction, magma erupted onto the seafloor in the form of pillow lava similar to that created at a mid-ocean rift. Eventually, the composition of the magma changed, and it began to erupt violently.

In its chamber, magma is not generally homogeneous or constant in composition. Instead, it is made up of many different elements and compounds. Some of these compounds, when melted, flow easily through cracks and vents in the Earth's surface. This is true of magmas containing iron, magnesium, and calcium. Other magmas, such as those containing potassium, sodium, and silica, are more viscous. Because of this, the lava is more resistant to flow. Instead, they will move into fractures in the rock and act as a plug that traps heated gases and rocks behind them. When the pressure becomes great enough, the lava explodes.

When magma erupts from its chamber, the first lavas to leave are those that flow easily. Basalt, which is rich in iron, magnesium, and calcium, is one of these lavas. What is left after the basaltic lavas leave are lavas that are richer in silicon, such as rhyolite and dacite, which does not flow as easily. These lavas tend to erupt explosively.

Geologists call the process by which a magma chamber is depleted of easily flowing basaltic lavas differentiation. By about 2.73 billion years ago, differentiation within local magma chambers set the stage for the eruption of rhyolite and dacite. Volcanoes began to erupt violently, producing thick beds of volcanic rock and debris. Local centers of volcanic activity began to develop, and shortly thereafter, volcanic cones began to protrude above sea level and form islands, such as the modern-day Philippines.

As soon as the volcanic cones poked above the surface of the water, they started to erode. Streams carried sediments, the products of erosion, into shallow seawater along the margins of the islands. Remember from the discussion of plate tectonics that colliding plates produce earthquakes. These earthquakes probably shook and loosened much of the sediments deposited along the margins of the islands, causing submarine landslides that transported sediments from shallow water near the island arc to the deeper water of the oceanic trench. As these swirling landslides of debris and water cascaded downward, the larger, heavier grains settled first and the smaller, finer grains last, creating graded beds of sand, silt, and mud called turbidites (fig. 11). Geologists like to find these beds because if they have been folded or tilted by later geologic processes, the beds tell which way was originally up.

Thus far, the story of the plates that eventually formed the park and the surrounding region had involved only one or two plates. A rifting plate splits in two, producing pillow lavas that erupted on the seafloor at the rift zone. Then the expanding plate collided with an adjacent plate and a chain of explosive volcanoes was born. Lavas from these volcanoes eventually built an island chain. Wind and water eroded the island chain and streams and earthquakes helped move eroded sediments, first into shallow waters surrounding the islands and then into a deeper trench near the place where the two plates were colliding. But there is much more to the story.

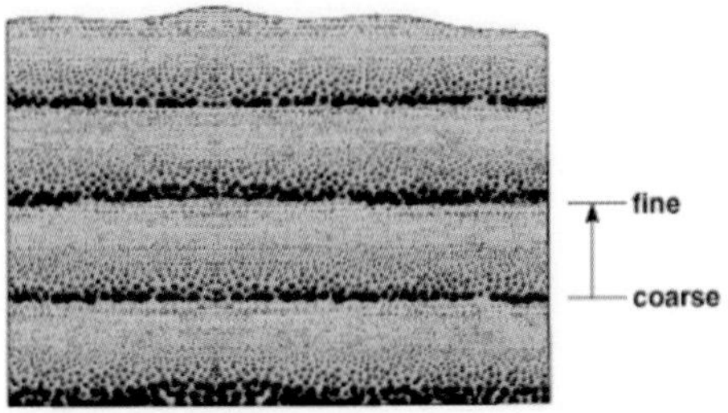

The geologic story of the region involves not one or two plates, but a series of plates, each with its own volcanic arc of islands and each colliding with the plate preceding it and the plate behind it (figs. 9a and 9b). To understand what this may have looked like, imagine you are standing on Steamboat Island on Rainy Lake (see park map insert on pages x-xi), except only it is approximately 2.7 billion years ago and you are standing not on a small island, but on a large island chain the size of Japan or the Philippines. As you gaze southward over Rainy Lake, you are looking across a vast ocean. The Kabetogama Peninsula is not visible. In fact, the sand, silt, and mud that eventually metamorphosed to form the present-day schist of the Kabetogama Peninsula are lying in muddy layers beneath the waves of this primordial ocean. Up and down the length of the chain are volcanic peaks wreathed in steam and the smell of sulfur fills your nostrils. There is no

Figure 11. Schematic diagram of graded beds showing coarse grains on the bottom and finer grains on top, indicative of sediment deposited in a submarine setting, USGS.

Figure 12a. Volcanic bomb, USGS.

Figure 12b. Lapilli tuff, USGS.

vegetation and the only sounds are those of the wind, water, thunder, and volcanoes. Welcome to the island chain that will eventually form the Wabigoon Belt (fig. 3b).

Looking southward, you may catch a glimpse of volcanic steam or the tip of the tallest volcano of a second volcanic island arc, at the very edge of the horizon, well behind the Wabigoon chain. If it were nighttime, the orange glow of erupting molten lava might light up the southern horizon. What you see in the distance is what will become the Wawa Belt (fig. 3b). From volcanoes on the Wawa Belt, booming volcanic explosions would reverberate over the waters, possibly followed by engulfing clouds of hot, sticky ash. Particles and fragments of volcanic rock would blast into the air and occasionally rain down, making strange satisfying sounds as they fall into the sea and equally strange but not so satisfying sounds as they pelt the ground around you. The larger particles are called volcanic bombs (fig. 12a) and run the gamut from about the size of your fist to that of a medium sedan. Particles smaller than your fist are called lapilli (fig. 12b). Anything smaller than about the size of a pinhead is known as volcanic ash (fig. 12c).

These two island arcs are moving toward each other. As they collide, all of the sedimentary deposits that had accumulated in the ocean basin between the two island arcs are being squeezed, squished, and folded. The squeezing and folding can be likened to two people pushing toward each other on opposite sides of a flat stack of rugs. The rugs represent layers of sedimentary rock. As the two people push, the region of carpet between them would buckle and compress, eventually forming a tightly compacted series of folds.

At the same time the sediments are being crushed, the subduction of many plates so close together causes widespread melting of the crust and

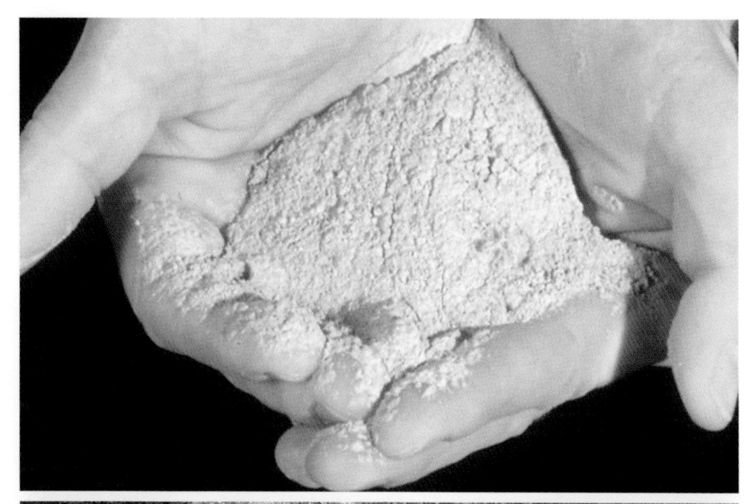

Figure 12c. Volcanic ash, USGS.

Figure 13. Biotite schist, Rainy Lake, NPS.

Figure 14. Migmatite outcrop on U.S. Highway 53 near Kabetogama, Minnesota, NPS.

the creation of huge magma chambers. The heat from these chambers combines with the intense compressional forces associated with collision to metamorphose the sediments into a class of metamorphic rock called biotite schist (fig. 13). Some of the magma also works its way up into the sediments and cools there, forming a spiderweb of veins, sills, and dikes. Evidence of metamorphosis and granite sills, dikes, and veins is found in a rock type called migmatite (fig. 14).

Chapter 7

PLATES COLLIDING AND MOUNTAINS RISING

Though island arcs are smaller than continents, they are made of the same material. Remember from our discussion of plate tectonics that when two pieces of continental crust collide, neither sinks below the other. Instead, they make mountains. The collision of many island arcs during the formation of the North American continent created several mountain chains, made of three types of rocks: volcanic, the backbone of the island arcs; the sedimentary rocks that had been squeezed between them; and deep below the surface, the granite bodies formed inside magma chambers that had cooled at great depth. All of these rocks were folded, faulted, and tilted skyward to form mountain chains that geologists believe were at least a few miles high.

2.6 BYA

Geologists call the building of mountain chains that accompanied the collision of island arcs in the Superior Province the Kenoran Orogeny. This orogeny is significant because, in the region of

Figure 15. Andes Mountains, South America, Wikipedia.

Voyageurs National Park, it is the oldest period of mountain building for which we have conclusive evidence.

Because there was still melted ocean crust under the colliding island crusts, it moved up through the overlying mountain chain and created many volcanoes. Geologists often cite the Andes in South America or the Cascade Range of the northwestern United States as good examples of what the volcanic mountain ranges that formed during this time may have looked like (fig. 15).

Near the end of the Kenoran Orogeny, the angle at which plates were colliding changed, which created a significant event in the geologic history of the northwestern part of the park. At this time, the Wabigoon and Wawa greenstone belts were sutured together, with the Quetico Belt sediments squashed between them. These three belts together formed the southern margin of the North American craton. They had been colliding with another plate farther to the south, like two cars meeting each other in a head-on collision. Between 2.6 and 2.7 billion years ago, the angle changed to an oblique one. If we use the car example, imagine one car coming from the north and heading directly south. The second car, or plate, instead of coming from the south, moves in from the southeast, heading northwest.

This change in the angle of plate collision had profound effects on the forces acting on the rocks. Changing the angle of collision caused the rocks not only to collide, but also to scrape past each other. To understand what happened, let us return to the chocolate and vanilla cake example. This time place your left palm around the upper left corner of the cake and your right palm around the lower right corner. Imagine that your left hand marks the north and your right hand

Figure 16. Migmatite along County Road 122, near Kabetogama Lake, NPS.

south. Slowly push your palms toward each other, while at the same time sliding your left hand to the right and your right hand to the left along lines parallel to the upper and lower edges of the cake. In other words, move your left hand toward the southeast and your right hand toward the northwest while sliding your hands past each other. What you will find is that the cake not only crumples between your hands but also begins to break apart along a line running down the middle of the cake parallel to the upper and lower edges. Eventually, pieces of cake on opposite sides of this line will scrape past each other.

A cake's material is fairly homogenous. If we mixed the batter thoroughly, there should not be much difference in texture and sweetness from one point to another. Crust at the point of plate collision is not as homogenous. It is composed of different rocks, each with a different hardness. When exposed to stress, rocks tend to break along planes of weakness. Most times, these planes of weakness are the boundaries between one type of rock and another. A great small-scale example of this phenomenon can be seen in migmatite. The dark rocks in the road cut are biotite schist and light rocks are granite. When road workers tore open the rock with explosives and machinery, they caused the rock to break. If you examine the road cut, you will see that many of the "breaks" or fractures happened where granite and schist meet (fig. 16).

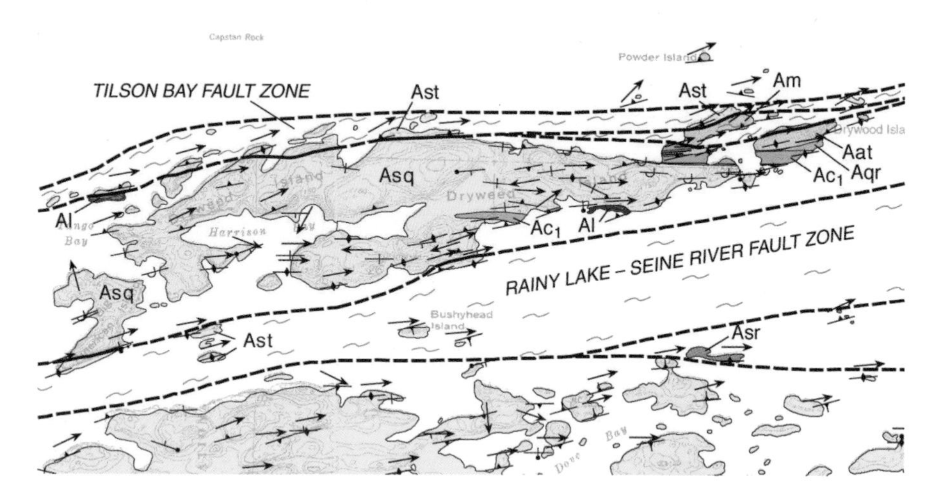

Figure 17a. Geologic map, west end section of Dryweed Island, Seine River Fault Zone, Rainy Lake, NPS.

Although the collision of two plates involved a zone of contact several hundred or even thousands of kilometers long, the rock still broke apart at the boundary between two different types of rock. In the park, rock broke apart at the contact between the volcanic rocks that made up the old island arc and the sedimentary rocks that were squished between island arcs. This zone of contact is now located in the northwestern corner of the park near Dryweed Island (fig. 17a). Geologists call this zone of contact the Rainy Lake–Seine River Fault Zone; it extends several hundred kilometers to the northeast into Canada. South of the fault zone, the rocks are mostly metamorphosed sediments and intrusive granite. Whereas north of the fault zone the rocks are mostly metamorphosed volcanic rocks (see geologic map).

During the past 110 years, the Rainy Lake–Seine River Fault Zone has attracted the attention and efforts of miners for gold and other minerals. A gold rush that began on Rainy Lake in 1893 centered on the fault zone. It is not uncommon to find rich

mineral deposits in Precambrian fault zones. By 1901 the Rainy Lake gold rush had ended. Many of the miners and their families stayed and settled the region (figs. 17b and 17c).

Figure 17c. Gold mine shaft showing the rocks layered due to the Rainy Lake–Seine River Fault Zone, Bushyhead Island, Rainy Lake, NPS.

One of the most interesting features created by the fault zone was something called a fault-bounded basin. Faults are not often straight lines but instead bend and wind in a way that may look like a child's drawing of a long snake. As the two sides of the snake-shaped fault slide past each other, a basin will open up. You can model this using only your hands: Take one hand and cup it inward slightly, place your other hand against it so that your fingers are bent backward. While your hands are held this way, slide your hands past each, keeping each hand rigid. A small space should open up in front of the cupped hand as your hands form sheer cliff walls that bind the basin. You have created a hand-bounded basin (figs. 18a and 18b)!

Imagine you are situated on the north wall of the basin and are facing south. Behind you rise sheer cliffs. Above and beyond the cliffs tower tall mountains created on the Wabigoon Belt. As you look to the south, you see a broad, fairly flat basin. Out in the basin, you see a single volcanic cone. The cone was created by the rising of subducted and melted continental crust that had been scraped off the shelf of the continent by a subducting ocean plate. The cone is draped with the hardened remains of individual lava flows and surrounded by a litter of rock and ash. It has been

Facing Page: Figure 17b. Gold miners, © Koochiching County Museum.

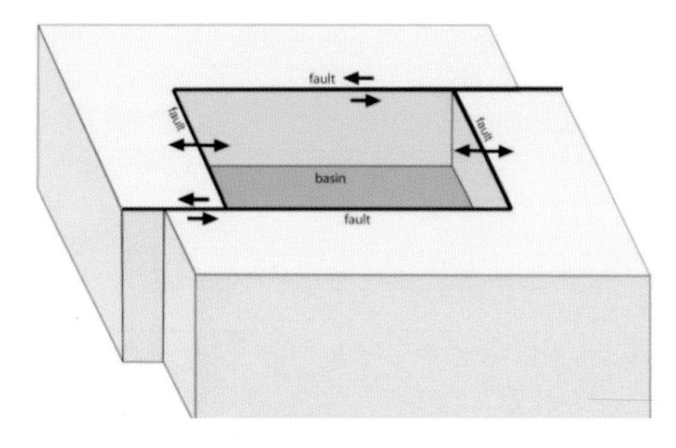

Figure 18a. Idealized image of what a fault-bounded basin might look like along a transform fault such as the Rainy Lake–Seine River Fault Zone, USGS.

carved by gullies and streaked with landslides. A large chunk of the western summit has been blown away. Though vaporous gases swirl up from the cone, the volcano is presently quiet and slowly eroding.

Looking north at the cliff face behind you, you notice that streams draining from distant mountains have carved gullies into the cliff face. The waters carry sediments—sand, silt, pebbles, cobbles, and even boulders—dropping some of them at the mouth of the gully. Over time, these sediments have piled up to form deep depositional aprons of rock called alluvial fans (fig. 19a). Finer sediments like silts and clays are carried out onto the basin floor to the south. After a severe flood, you may see something called a playa lake in the flat belly of the basin (fig. 19b). Playa lakes are lakes that form in an enclosed place, where

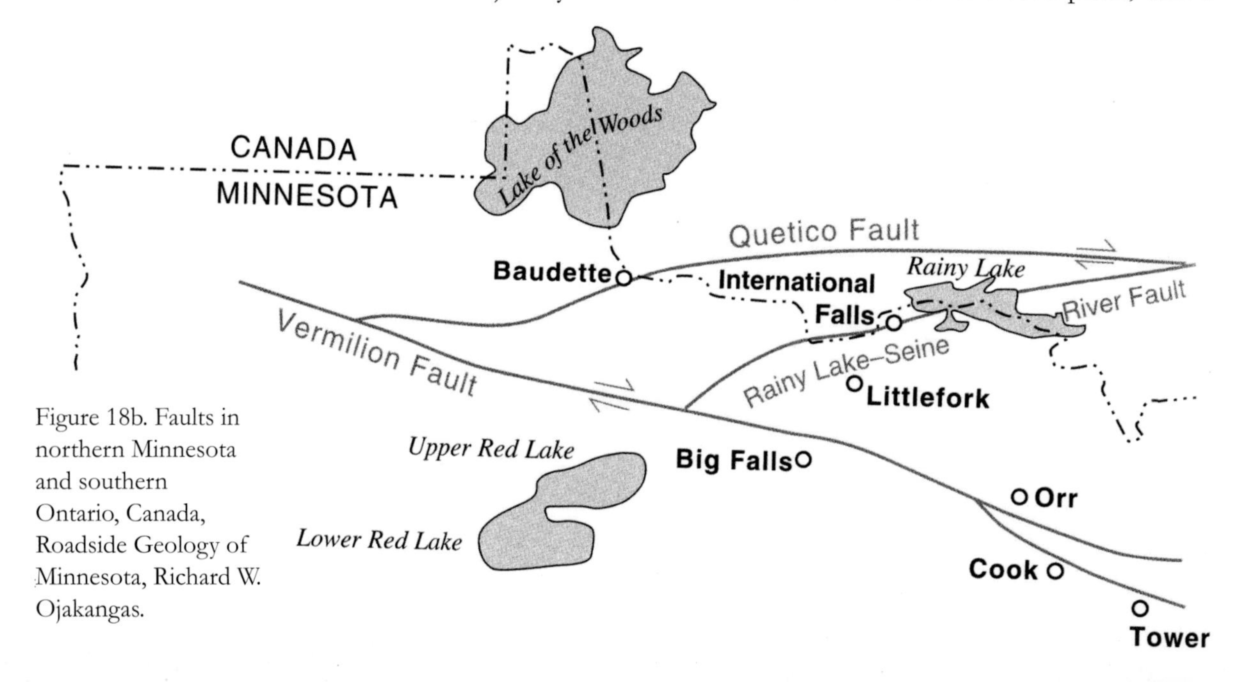

Figure 18b. Faults in northern Minnesota and southern Ontario, Canada, Roadside Geology of Minnesota, Richard W. Ojakangas.

water can leave only through evaporation.

If you stand in this place for tens of thousands of years, you will notice that earthquakes are common. Each new quake subtly affects the landscape. After a quake, the basin walls may be higher and located a little farther to the right.

The fault-bounded basin we described above is pertinent to our discussion because we can find evidence of it in rocks within the park. Using the Geologic Map of Voyageurs National Park, locate Dryweed and Drywood islands on Rainy Lake. On these islands, one can find rocks that provide clues to understanding the first ten to hundred million years following the development of the Rainy Lake–Seine River Fault Zone and a fault-bounded basin. Collectively, these rocks are called the Seine River Series.

Figure 19a. Birds-eye view of a modern alluvial fan in Death Valley National Park. The alluvial fans forming along the margins of the Archean fault-bounded basins in Voyageurs National Park would have looked very similar, NPS.

Figure 19b. Cross section through an alluvial fan in Death Valley National Park showing rounded boulders, cobbles, pebbles, sand, and a playa lake basin in the background, NPS.

Chapter 8

THE SEINE RIVER ROCKS

2.5 BYA

The north edge of Dryweed Island is highly sheared (which indicates that it was near the north edge of the fault zone; the collision of plates sheared the rock). It is composed of basaltic greenstone (fig. 20). It probably contains pillows, but the rock is too highly sheared to show such features. This basalt most likely formed the platform upon which the volcanic island arcs were built. Erosion of the islands may have helped expose the basalt, and the collision of plates probably folded the rock and pushed it upward so that it formed part of the floor of the basin.

Traveling across the island to the south, we next find volcanic rock geologists call accretionary lapilli tuff (fig. 21). Its source was probably a volcano on the floor of the fault-bounded basin. You can see that the rock looks like a pile of ravioli stacked on top of one another. All of the ravioli are roughly the same in shape and size. How did this rock form? When volcanoes erupt, they spray huge amounts of fine ash thousands of meters into the atmosphere. Sometimes, these ash particles will coalesce around a water droplet or a small piece of dust. This is thought to happen when rain or steam flows through the ash cloud of an eruption column. The water or steam wets the ash particles and makes them sticky. Then they begin to attach to other ash particles, much in the same way a ball of dough forms when a drop of water is added to flour. These globs of sticky volcanic ash grow until they are anywhere from a fraction of an inch to a few inches in size. All of this is happening in the atmosphere as the ash is being blown skyward by the force of the volcano and then slowly falling to Earth. The force of wind on the descending globs of ash rounds them into balls.

Figure 20. Greenstone schist, Skipper Rock Island, Rainy Lake, NPS.

If you have the chance to examine this outcrop, you may notice another feature that offers some clues about the ancient basin. Volcanoes can explode with terrible force, sometimes cutting U-shaped channels or valleys through a deposit of fresh-fallen lapilli tuff. An example of a U-shaped channel can be seen in figure 22. Like all channels and valleys, the bottom of the "U" will always lie on the earthward side. This U-shaped valley offers the first piece of evidence that the rocks we are examining have been

Figure 21. Lapilli tuff, east Dryweed Island, Rainy Lake, NPS.

Figure 22. U-shaped channel, www.sciencedirect.com.

tilted on their sides. This lapilli tuff originally lay flat on the bottom of the ancient basin. Since that time, the forces of plate collision have tilted it upward. The U-shaped channels also tell us that the top or younger portion of this series of rocks lies to the south. This makes sense because we know that the pillow lavas, the oldest rocks in the park, lie to the north.

This outcrop of lapilli tuff also helps us re-create a piece of the basin landscape. Lapilli tuff probably formed part of the litter of rock, ash, and lava flows that encircle the cone. Lapilli tuff and U-shaped valleys are only formed during very violent volcanic explosions, those strong enough to tear off the side of a cone. This evidence of violent explosion is the reason it was suggested that the cone in the valley had part of its western summit missing.

If you travel south from the lapilli tuff, you will notice overlapping beds of two other types of volcanic rock: a light green-gray rock called rhyolite and a dark green rock called mafic tuff (fig. 23). Because these rocks are located to the south, we know that they are younger than the lapilli tuff and were probably deposited on top of it.

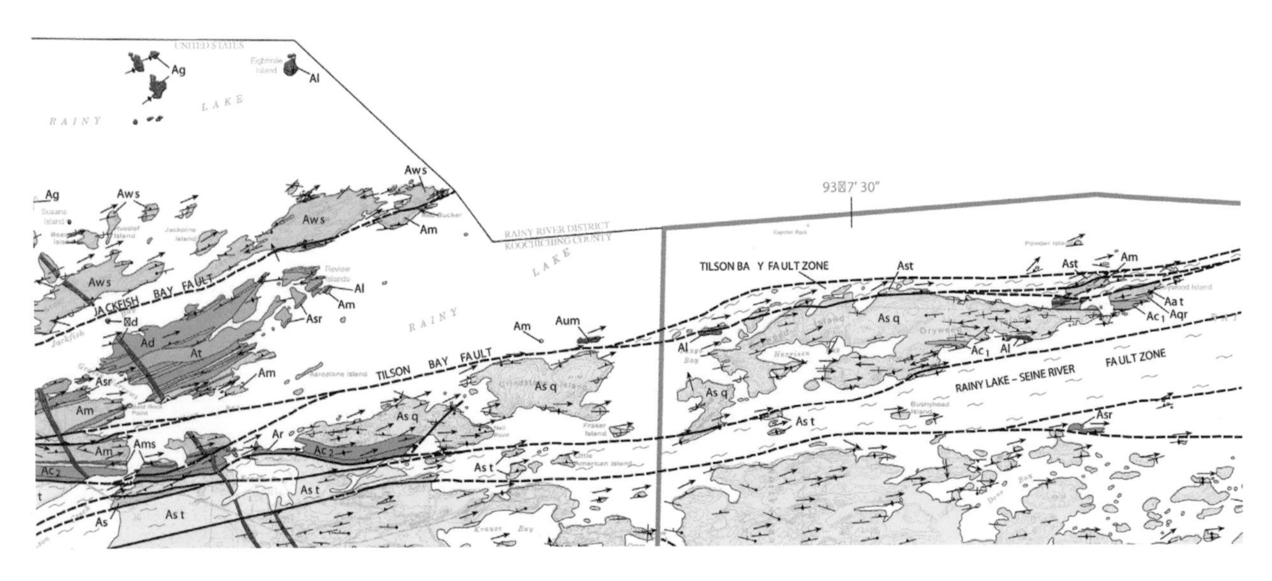

Figure 23. Map of mafic rocks located in the park, NPS.

If you pick up a chunk of this rock and look closely, you may be able to see tiny, iridescent, blue quartz crystals, called quartz-eyes (fig. 24). How was this rock formed? Many times after a violent explosion like that which produced the lapilli tuff, lava will pour out of cracks and fissures in a volcano and cover everything in

its path. If you return to the volcanic cone mentioned above, imagine that through a veil of acrid clouds of steam and ash, streams of red-hot, viscous, liquid lava are pouring out of fractures all around the cone, like a serving of gravy flowing over a pile of mashed potatoes. Individual flows slowly cover portions of the accretionary lapilli tuff deposits. A quick look at the park's geologic map shows that this rock type pinches out before it reaches the western end of Drywood Island on Rainy Lake, suggesting that the lava flow made it only that far.

Figure 24. Example of blue quartz-eye that can be found on Dryweed Island, Rainy Lake, NPS.

The rhyolite and lapilli tuff we see in the park were probably not created by a single volcanic event. Instead, a series of eruptions probably formed the lapilli tuff, and a series of lava flows formed the rhyolite. Fine volcanic debris, such as lapilli and ash, is easily eroded soon after an eruption. In the time between volcanic events, wind and water actively eroded volcanic sand and ash before it had a chance to be buried by younger deposits of lapilli or lava. Wind and water may deposit some of these eroded volcanic sediments into beds. Geologists call these deposits of eroded ash and sand mafic tuff. As mentioned above, it is interbedded within the rhyolite, indicating that eruptions alternated with periods of erosion.

The next rock type to the south after the rhyolite and mafic tuff is called polymictic conglomerate (fig. 25). The rock appears to be composed of cigar-shaped pieces of many sizes enclosed in a matrix of gray-green rock. What is this rock made of and how did it form?

Figure 25. Metaconglomerate, Gold Shores Drive, International Falls, Minnesota, NPS.

Polymictic conglomerate is just a technical term for a rock that is a collage, a conglomeration of smaller rocks held together in a matrix of natural cement. It is a sedimentary rock, which means it formed from materials that were carried from elsewhere. Where do we find deposits of smaller rocks in the shapes and sizes found in polymictic conglomerate? We usually find them in streams draining steep terrain. In fact, the gravel and pebbles in this rock were probably deposited in an alluvial fan. Alluvial fans are commonly found along valley walls in many places around the world, such as Death Valley in California. They form where fast moving streams, carrying material eroded from highlands above the valley, rapidly slow when they reach the valley floor. When streams slow, the water no longer has the force needed to carry gravel, pebbles, and even some sand. It drops these sediments, and they pile up into deposits resembling fans or a single clamshell placed open end down. Eventually, the sediments are buried deeply and cemented together into a conglomerate. Why is it probable that the gravel that came to form this conglomerate was deposited in an alluvial fan? If you look at the edges of the gravel and pebbles in the conglomerate, they are rounded. They have been scraped and abraded and worn by their trip downstream and into the basin. The rivers and streams that carried this sediment were those we described above. They were probably found at the mouth of large gullies in the cliff faces that bounded the basin.

You can learn a few more things from this conglomerate. First, you may notice that the pieces of rock in the matrix are oblong in shape. They look like stream gravel that has been flattened, and they are. They were probably flattened by the continued collision of the Wawa and Wabigoon plates with a plate farther south. The fact that the rocks have been compressed tells us that the plates were still colliding after the rocks were deposited. You can model this process by squeezing a round ball of clay in your hand. The clay will flatten into a cigar-shaped object, the ends of which squeeze out the two openings between your palm and fingers.

You may also notice that some of the oblong gravel and cobbles in the conglomerate twinkle with flecks of iridescent blue. The blue comes from tiny blue quartz crystals embedded within the gravel and cobbles. If you look closely at an outcrop of conglomerate, you will notice quite a few small pieces of quartz-eye in it. In fact, a rough count reveals that they are the dominant types of gravel or pebble in the polymictic conglomerate. These are sediments eroded from the blue quartz-eye rhyolite we encountered in the volcanic flow just to the north! Their presence suggests that erosion of the rhyolite volcanic flows was occurring at the same time that other sediments were pouring into the basin. It also tells us that the rhyolite is older than the conglomerate. Erosion and weathering of the rhyolite volcanic flows produced pebbles, cobbles, and boulders, which were incorporated into alluvial fan deposits along the margins of the Rainy Lake–Seine River basin.

Figure 26. Buff-colored metaquartzite, Dryweed Island, Rainy Lake, NPS.

Now let us head west and search the shoreline of Dryweed Island, just across the small inlet between Drywood and Dryweed islands. We soon locate the conglomerate unit again and find that its southern margin is in contact with a yellow-tan metamorphic rock called metaquartzite, shown in figure 26. Some of the wave-washed points along the southern shoreline of Dryweed Island provide excellent examples of this rock. How did this rock form? The material that composed this rock was originally deposited as sand in a river system that flowed from the base of the alluvial fan deposits. How do we know this? Careful examination reveals important structures called crossbeds, which are the remains of migrating ripple marks that form in sandy stream bottoms. If you have been in a sandy stream bottom draining steep terrain, you know that it is not flat but is instead made up of many small ridges, like a washboard. The stream deposits sand grains on the downstream slope of each ripple in the washboard. Over time, the sand grains form layers parallel to the slope of the ripple.

Because they preserve the slope of ancient ripples in a stream bottom, crossbeds can tell us which way was originally up and which way the current of the stream was flowing when it formed (fig. 27). Because the crossbeds slope to the north and west, we know that the top of the rock unit is to the south (consistent with our age relationship on Drywood Island) and that the dominant current direction was to the west.

Figure 27. A set of crossbeds with orange-colored glacial striations running diagonally across the photo, Dryweed Island, Rainy Lake, NPS.

This ends our tour of the ancient Rainy Lake–Seine River basin. We are now standing on the southern edge of Dryweed Island. Over the next ten million years or so following the creation of the crossbeds, the Wawa and Wabigoon plates will continue to collide with the plate to the south. The evidence presented above in the lapilli tuff and the polymictic conglomerate tells us that the force of the colliding plates will compress the rock, folding the layers up accordion style. Forces of erosion, in time, will saw off the "noses" of those folds, leaving behind the tilted layers of rock we see today.

Chapter 9

THE RAINY LAKE–SEINE RIVER FAULT ZONE

The Rainy Lake–Seine River Fault Zone is what some geologists refer to as a wrench zone (fig. 17a). The basin occupied a triangle between two sides of the fault. Previously, it was indicated that the plates moving past each other at a winding fault or contact zone opened the basin. As the plates continued to move, they compressed the basin and wrenched it to the right. The rocks in the basin began to fold under the force of compression. Because the plates were sliding past each other at the same time, the folds were twisted so that their axes ended up pointing downward, or plunging to the northeast.

The forces of compression also helped metamorphose rocks in the wrench zone. Under intense compression, minerals in rocks rearranged to reduce the stress on them. Minerals that were polygons may reform themselves as flattened shapes. This is called foliation and is evident in the vertically oriented

Figure 28. Locator Lake Fault Zone, Chain of Lakes, Kabetogama Peninsula, NPS.

slabs of rock throughout the park. As these minerals are squeezed by transgression they elongate in such a way that the long axes plunge downward to the northeast in the same orientation as the fold axes. It shows some of these plunging lineations, which are fantastically displayed on the vertical slabs of rock, called vertical foliation planes, throughout the park's metasedimentary and metavolcanic rock. All of this occurs at depth in rocks that are metamorphosing and deforming in somewhat of a plastic fashion.

Sometime after 2.6 billion years ago the plates stopped moving. Initially, small-scale crustal adjustments probably continued to take place in response to cooling of the crust and slow exhumation. One possible result of these crustal adjustments is the formation of numerous fractures or faults. One of these late-stage faults, the Locator Lake Fault, can be readily observed underlying the Chain of Lakes on the Kabetogama Peninsula (fig. 28). These faults and fractures define weak zones in the rock that are easily eroded and weathered. Preferential erosion and excavation of these weak zones created low-lying regions through which water could flow and eventually influence the location of the lakes.

Figure 29a. A portion of diabase, gabbro, and andesite dike near Kabetogama Lake. Located along County Road 122, Kabetogama, Minnesota, NPS.

Sometime around 2.1 billion years ago, the crust throughout northern Minnesota and northwestern Ontario fractured and pulled apart. Long, thin crevices opened in the bedrock and into these crevices seeped magma. The magma cooled and is now present as hundreds of thin elongated units of rock (150–400 feet thick and several miles long) that conspicuously cut across older rock units. Geologists call a line of rock that crosscuts older rock a dike, and a collection of many dikes is a dike swarm. The dikes in the vicinity of the park are northwest trending vertical slabs of diabase, gabbro, and andesite—very hard, dark gray to black rocks, that are finely to coarsely crystalline in texture (fig. 29a).

Figure 29b. Part of a dike swarm on County Road 122, three miles north of U.S. Highway 53, Kabetogama, Minnesota, NPS.

A major dike from this swarm cuts through the park on the Echo Bay Trail and is best exposed on County Road 122 near the Kabetogama Lake Visitor Center (fig. 29b). Although geologists are not sure what caused the fracturing, it was undoubtedly related to tectonic processes, such as continental collisions along the margins of the North American plate. This dike and others like it are the youngest exposed bedrock in Voyageurs National Park.

After the dike swarm, erosion wore down the mountains, exhuming and exposing ancient, frozen magma chambers, sequences of volcanic sedimentary rocks, and the vertical to near-vertical limbs of the folds.

Chapter 10

FROM MOUNTAIN PEAKS TO MOUNTAIN ROOTS: THE MISSING TWO BILLION YEARS

After the dike swarm 2.1 billion years ago, the process of new rock formation at Voyageurs National Park goes strangely silent. The volcanoes are extinct, and earthquakes are rare. This part of the North American continent has formed. Plate collisions still take place and will continue to for the next two billion years but probably not at Voyageurs. They will occur along the now distant continental margins, so far away that there is little evidence of them in the park's rocks. Here the geologic story seems to abruptly stop. It will not resume until the glaciers move across the continent during the Pleistocene. You can see this missing period 2.1 billion years through 1.6 million years in (fig. 2).

The face of places,
and their forms decay;
And that is solid earth,
that once was sea;
Seas, in their turn,
retreating from the shore,
Make solid land,
what ocean was before.
Metamorphoses, XV – Ovid

1.6 BYA

When we say the geologic story seems to stop, we mean there is no evidence in the rocks of events after that time. We do know that two billion years ago the landscape of the park was a rugged mountainous highland. We can safely surmise that the forces shaping the landscape from that time until quite recently were subtle: wind, water, and ice. These forces slowly chewed through layers of solid rock, gradually eroding the mountains and uncovering layers of rock, the roots of the mountains, which lay kilometers below the peaks. Erosion removes pages from the geologic history book. There were undoubtedly streams and rivers, and perhaps lakes in the region during the past two billion years. But what exactly existed in the region is anyone's guess.

Though the rock record in the park is silent for about two billion years, regions to the east, west, and south are rich with evidence of events both geologic and biological. Geologists have used this evidence to infer what was happening in the region of the Canadian Shield that includes the park.

Life evolved in spectacular ways during the two billion years of missing history. As we mentioned early in the last chapter, the first living organisms appeared prior to the plate collisions that formed the core of the North American continent. These were single-celled prokaryotes similar to some modern bacteria and blue-green algae with evidence indicating that they evolved more than 3.5 billion years ago (although new evidence continues to push the origin of life back further and further). Eukaryote cells contain a nucleus and make up the bodies of other single-celled organisms and all animals, plants, and fungi. Most biologists feel they had evolved by at least one billion years ago. There is no evidence of animal, plant, and fungi fossils in the park's rocks, mainly because the rocks formed prior to the evolution of multicellular life-forms.

Most evidence of early multicellular life is preserved in marine sediments that piled up in warm, shallow, Mediterranean-type seas. Sedimentary evidence for these warm, shallow seas is present in the Minnesota and Canadian region close to the park. Is it possible that this region was covered by these oceans, and that all of the resulting marine sediments have been eroded away? Because Voyageurs' landscape does not exhibit these deposits, it is generally assumed that the park was never completely covered by such shallow seas.

Our description of the park, throughout the last chapter, was one of a bare, rocky, volcanically violent landscape. Because the life-forms we described above evolved in warm, shallow seas, they probably did little to change this scene. Eventually, plants would colonize the land. The earliest plants yet found with vascular tissue to carry water and nutrients date to 425 million years ago. Many scientists think the land was sparsely occupied by lichens (in dry places) and by mats of algae (in wet places) before this. Animals, arthropods (ancestors of modern insects), and then lungfish are thought to have followed plants onto land. By three hundred million years ago, primitive forests of spore-bearing plants, large amphibians, arthropods, and early reptiles probably occupied the land in the region of the park. As with evidence of marine life, there is no evidence within Voyageurs National Park of terrestrial plants or animals prior to about 12,000 years ago, but rest assured that they were here. It is almost certain that

425 MYA

creatures lived and died on the slowly eroding rocks of the park. Their remains were most likely wiped away by wind, water, and ice.

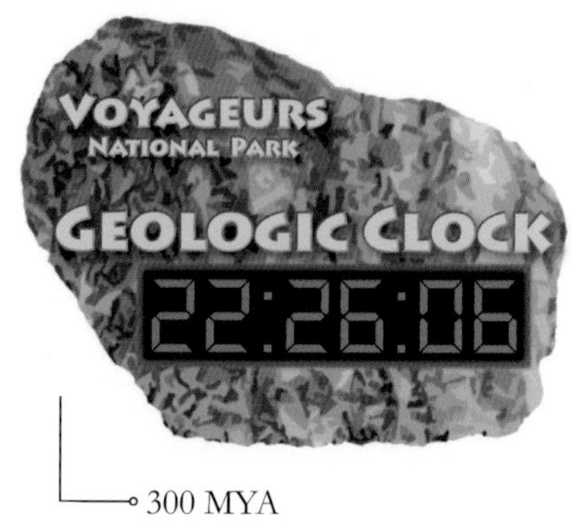

300 MYA

Though we may not know what life or the landscape of the Voyageurs region looked like, we do know where on the face of the planet the area was located. As was discussed in Chapter 2, plates move, carrying their respective continents and landmasses with them. We can reconstruct the movement of continents using a tool called paleomagnetics (fig. 30). This is how it works: As magma cools, minerals containing iron align themselves with the Earth's magnetic field (the iron in a compass does exactly the same things). Once the magma is solid, it preserves an ancient arrow pointing in the direction of ancient poles. As the continents move, the rock's position may twist and turn, but it will keep that arrow. By looking at a host of rocks of the same age from all over the world, we can triangulate the position of the poles and reconstruct the ancient positions of the continents.

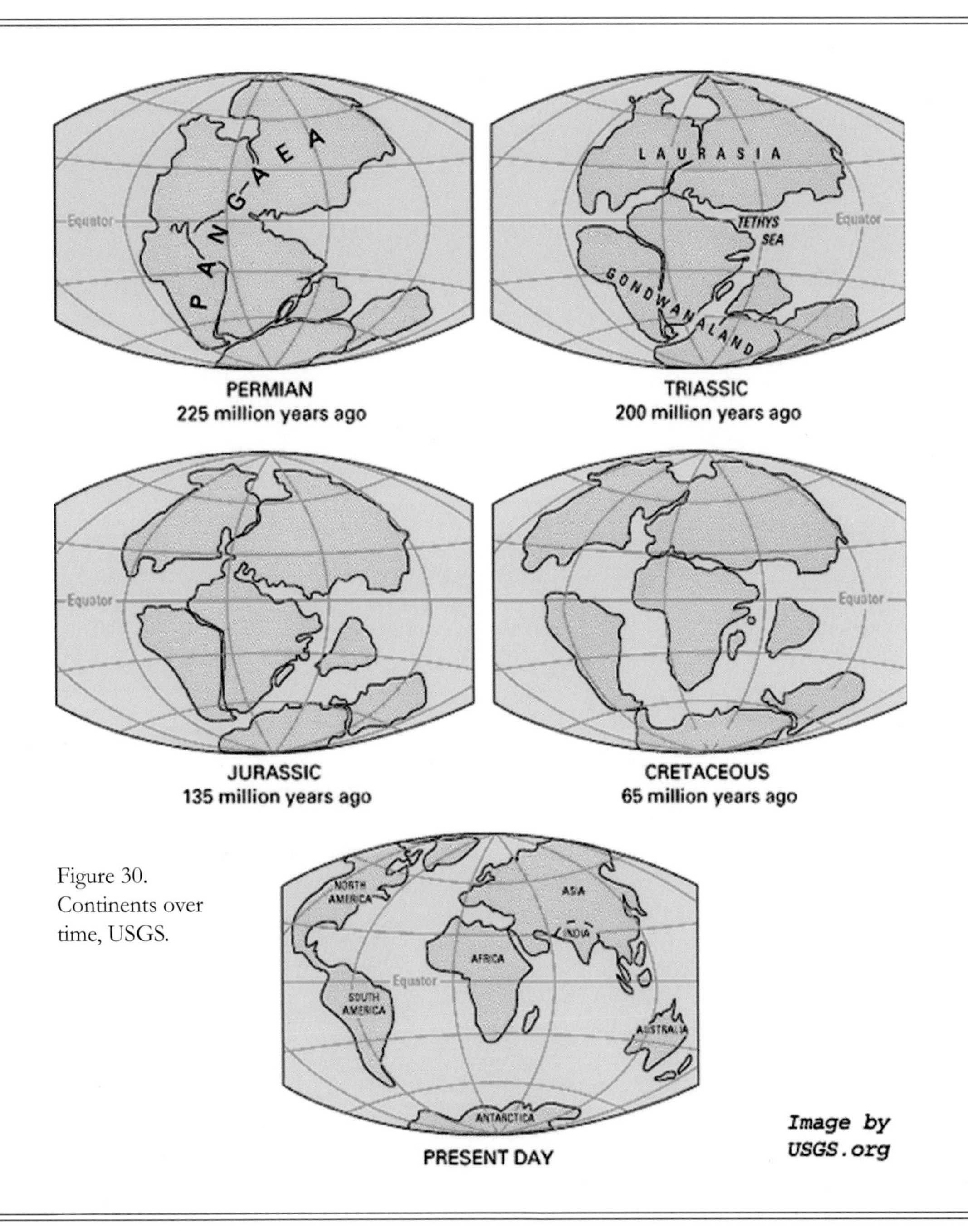

Figure 30. Continents over time, USGS.

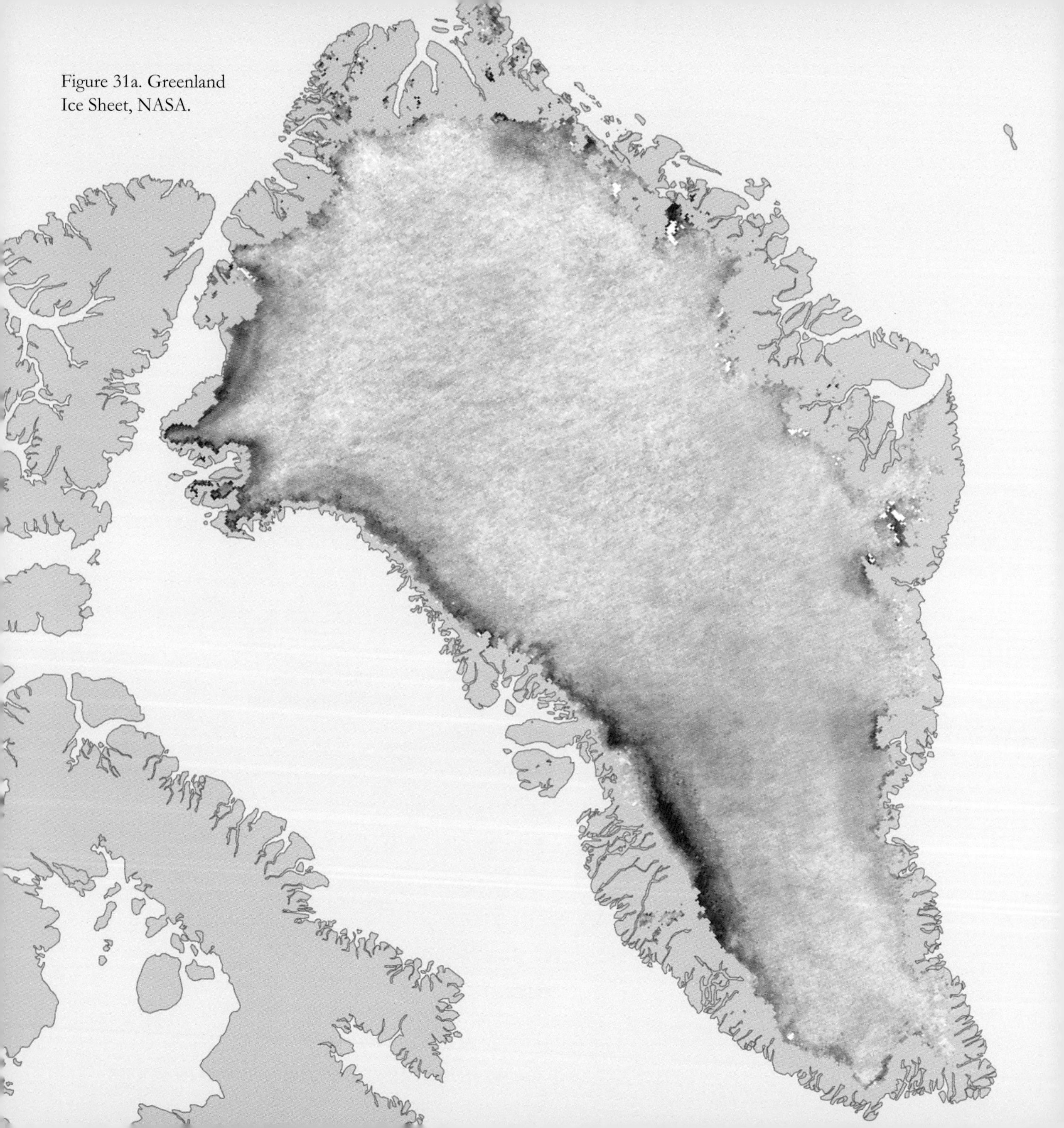

Figure 31a. Greenland Ice Sheet, NASA.

Chapter 11

PLEISTOCENE GLACIERS

It is a January evening in Voyageurs National Park. The temperature is minus thirty degrees Fahrenheit, and the ice on the lakes is three feet thick. The landscape seems desolate and empty. It is difficult to imagine the arrival of warm weather, but summer will come. Warmer temperatures will melt the ice, and in five months, on a July afternoon, hikers will be bushwhacking through a hot forest to reach a cool lake for a swim. But there was a time, starting more than 40,000 years ago, when summer temperatures were not high enough to melt thick sheets of ice that covered the landscape.

> *...Here was a plastic, moving, semi-solid mass, obliterating life, swallowing rocks and islands, and ploughing its way with irresistible march through the crust of an investing sea.*
>
> Journal entry describing the Humboldt Glacier, Greenland, Elisha Kent Kane, April 27, 1854

The Great Ice Age, as it is commonly called, began about two million years ago and lasted until about 12,000 years ago. During the Ice Age, great sheets of ice, called continental glaciers, periodically covered nearly all of Canada and most of the northern United States. At their thickest, they were probably two miles from top to bottom. Flowing outward from their birthing grounds in northern Canada, some of these massive ice sheets

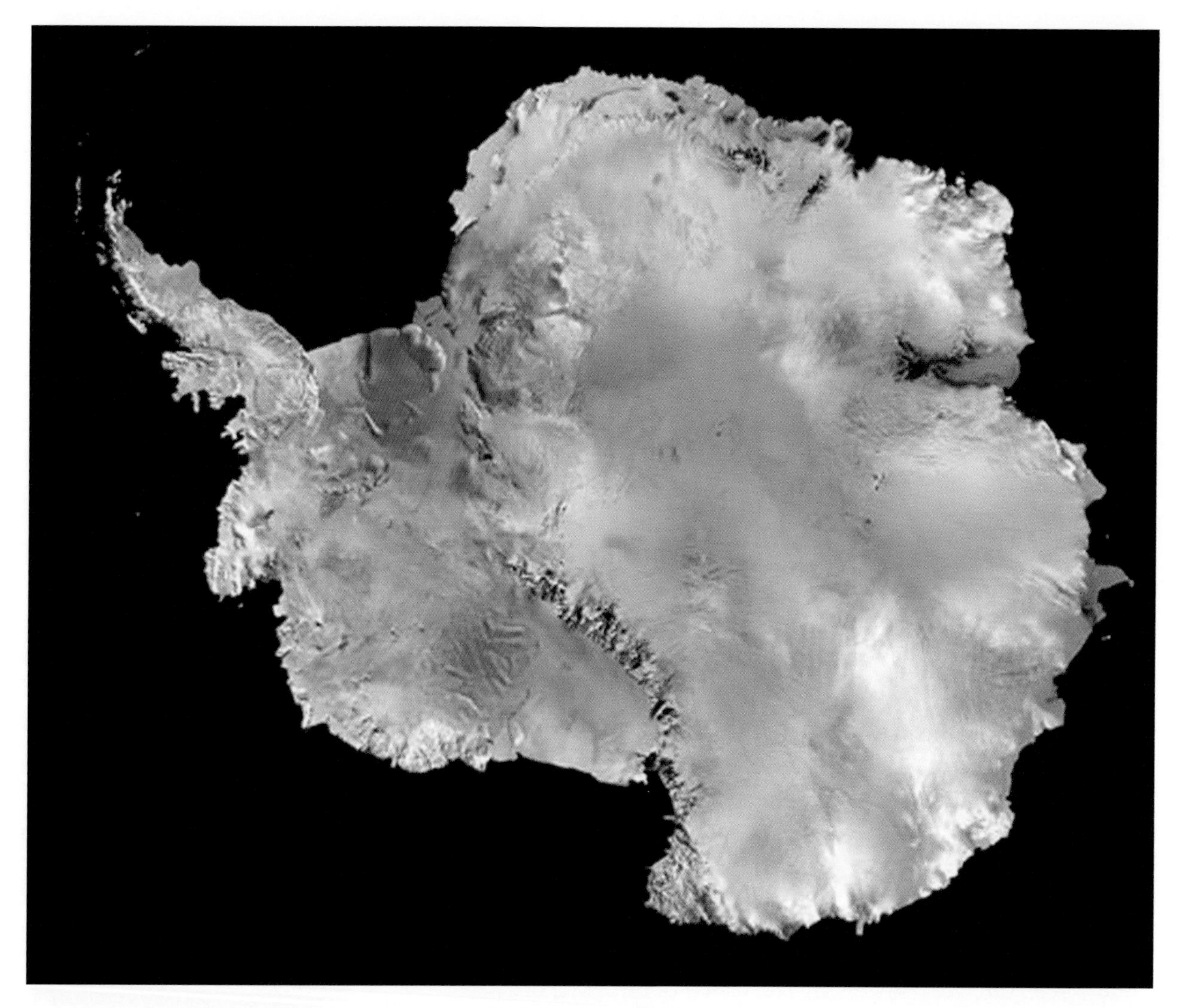

Figure 31b. Antarctic Ice Sheet, USCG.

ground their way as far south as southern Illinois. You can gain an appreciation for what these continental glaciers looked like by examining photos of large ice sheets in places such as Antarctica and Greenland (figs. 31a and 31b).

For those who are interested in geologic terms, geologists call the time that included the Great Ice Age the Pleistocene epoch. It constitutes the first part of the Quaternary period (fig. 2).

You may have noticed an inconsistency in the periods of time cited above for the Ice Age. It was stated that ice covered the park at least 40,000 years ago. Yet it was also stated that the Great Ice Age, or the Pleistocene, began two million years ago. What was happening in the park between two million and 40,000 years ago? The answer is that the park's rocks do not tell us. As will be discussed later in this chapter, glaciers swept clean much of the surficial geologic record. The glaciers that arrived about 40,000 years ago swept away any evidence that may have been left by earlier glaciers. But geologists, like all scientists, are resourceful. If they cannot find evidence in one place, using one method, they look for it in other places using a new approach.

Geologists have found evidence of earlier glaciers in southern Minnesota, Wisconsin, Illinois, Iowa, Kansas, and Nebraska. Since these glaciers flowed to places south of the park, we can assume that they also covered this area. There is evidence of at least three periods of glaciation prior to the last glaciation. Each glacial period was separated from the next by an interglacial period, when the climate warmed, the glaciers melted, and flora and fauna returned.

Glacial geologists have also uncovered evidence of earlier glaciation in a most unlikely place—the bottom of the ocean. What were glacial geologists doing there? The amount of water on the Earth's surface and in the atmosphere is fairly constant. During

40,000 YA

an ice age, much of the planet's water gets locked up in glaciers, and the levels of oceans can fall by a few hundred meters.

Many people may be surprised to learn that not all oxygen or water molecules are the same. You may have heard the word *isotope.* Most elements on our planet, including oxygen, exist as a number of different isotopes. The most common isotope of oxygen has eight protons and eight neutrons. It is called O_{16}, for eight neutrons plus eight protons. Another isotope of oxygen, much less common, is O_{18}, which has eight protons and ten neutrons. Each of these isotopes behaves a little differently. For example, O_{18} is heavier than O_{16}. Both isotopes can combine with two atoms of hydrogen to form a water molecule. When they do this, the water molecule with O_{16} evaporates more easily than that with O_{18}. This means that the water that eventually forms the snow that falls on glaciers will be enriched in O_{16} compared to seawater. Over time, as the glaciers build, the water that remains in the ocean will lose O_{16} and become more and more concentrated in O_{18}.

If geologists could find some record of what the O_{16} to O_{18} ratio was in the past, they could tell us whether there were glaciers present at that time. Luckily, small animals called foraminifera use the oxygen dissolved in seawater to make their shells. These creatures live at the ocean's surface. After they die, they sink to the bottom. The enterprising geologist can take cores of the ocean bottom, retrieve these shells, and determine the ratio of O_{16} to O_{18} in earlier times. Using evidence from foraminifera, geologists suggest that there may have been as many as ten glacial periods in the past million years, six more than we have evidence for on the ground.

As was stated above, the early glacial chronology of the park is not known with certainty because the signs of older glaciations have been obscured by erosion. What we do know is that about 75,000 years ago the climate began to cool, and the most recent period, called the Wisconsin glaciation, started. The Wisconsin lasted from 75,000 to 12,000 years ago. Ice first started to accumulate in a region centered on Hudson Bay. Expanding eastward and westward across Canada, it eventually coalesced with other regional glaciers to form a massive continental glacier, known as the Laurentide ice sheet (fig. 32). The oldest dated glacial evidence from the Wisconsin suggests that by at least 40,000 years ago Minnesota, including the area now known as Voyageurs National Park, was covered with ice.

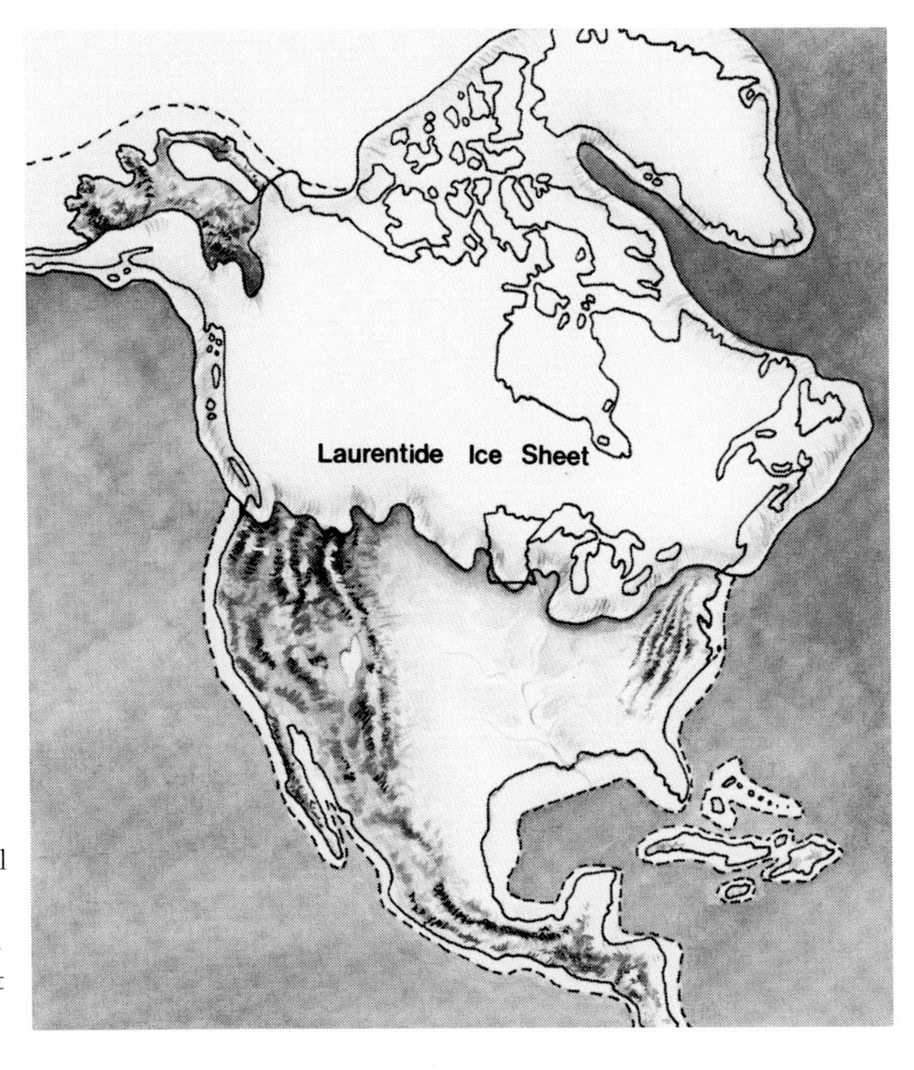

Figure 32. Laurentide Ice Sheet, *Roadside Geology of Minnesota*, R. W. Ojakangas.

Geologists define a mineral as a naturally occurring solid, with an ordered crystal structure and a fixed chemical composition. Ice satisfies these conditions; therefore, it is a mineral. It is a solid below thirty-two degrees Fahrenheit. It crystallizes in ordered patterns, as we witness with snowflakes. And it is made up of water molecules. It is different from most minerals we are familiar with in that it melts at a very low temperature, temperatures that we judge as comfortable. By comparison, the minerals in ultramafic rocks will not melt until they are heated to 2,460 degrees Fahrenheit.

Snow and ice will build a glacier only under very special conditions. First, the mean global temperature of the earth must be cool for a prolonged period of time, about eight degrees Fahrenheit cooler than today. Under these conditions, summer temperatures are cool enough for some of the winter snow to survive until the next winter season, and over time, layers of snow accumulate.

When snowflakes first fall, they are pointed in shapes similar to grade-school cutouts. As the season wears on, these flakes will lose their points and become the granular snow skiers call corn snow. If buried deep enough in a snowbank, granules may partially melt and refreeze into aggregates of granular ice crystals called firn. This occurs quickly, often within one season. Firn looks like small (1–2 mm in diameter) clumps of hardened snow. Firn is to snow as sandstone is to sand. As snow continues to fall, the weight on top of the firn will crush the firn clumps together, squeezing out air. Some of the firn crystals may melt and re-freeze to form cement. Crushed, tightly packed, and cemented firn is glacial ice. And it is only at this point, when the ice is thick enough to be reformed (or deformed) under its own weight that a glacier is born. After 30,000 to 40,000 seasons, enough glacial ice can form to cover one-third of the continent.

When ice is under enough pressure, say within a mile-thick glacier, it also flows. To flow, individual ice crystals within a glacier must slide past one another. They do this either by deforming (compacting or stretching) or by melting and refreezing. As you can imagine, these small events take a long time to add up. But when they do, glaciers move. Because ice is very stiff, it flows quite slowly. Sometimes glaciers may move faster by sliding on a thin skim of melted water produced at the base of the ice.

Namakan Lake entering the dam at Kettle Falls, NPS.

In what direction do glaciers move? Like water or piles of sand grains, glaciers flow downhill under the force of gravity. How then did the continental glaciers flow from Hudson Bay, which is close to sea level? The answer is that the glaciers flowed like a pile of sand or salt. At their center over Hudson Bay, they built up until they were perhaps more than three miles thick. The ice on top was then higher than terrain to the south and flowed out over it.

The annual accumulation of new snow each year determines whether a glacier advances or retreats. Glaciers retreat when more snow and glacial ice is being lost through melting and evaporation than is being added. The term *glacial retreat* is misleading—glaciers do not normally reverse their flow. Instead, flow stops near the terminus of the ice; the ice stagnates and begins to melt in place.

Glaciers and ice sheets take much less time to melt than they do to form. For example, the Wisconsin glaciation began about 75,000 years ago. Its ice lobes were advancing until about 12,000 years ago. By 12,000 years ago, glacial ice began to melt in the park. By 8,000 years ago the southern margin of the glacier had retreated to Hudson Bay. Thus, it took 63,000 years to build the glacier to its maximum and only 6,000 years to destroy it!

Glaciers reshape the landscape by eroding the rocks and soil in their path and carrying this mishmash of material until it is disgorged in huge trash heaps of sediment. Often, glacial ice is so full of sediment that it looks black. In this way, glaciers are like big conveyor belts or snowplows, carrying material as far as hundreds of miles from its point of origin.

Glaciers erode rocks by processes called abrasion and plucking. In abrasion, rock debris freezes into the bottom of the glacier and acts like coarse sandpaper. Large rock fragments produce striations in the bedrock, allowing us to deduce the regional flow patterns. The striations in figure 32 are oriented to the northeast at about thirty degrees, indicating the direction from where the ice came. More finely crushed rock produces a smooth, polished surface. Evidence of both types of abrasion, polishing and striations, is common throughout the park. Areas of the shore kept bare by waves and ice provide the best exposures.

Figure 32. Glacial striations, polishing, and chatter marks, Frank Bay, Rainy Lake, NPS.

While abrasion can scour rocks and scrape off a thin layer of fine materials, plucking is a glacier's most effective tool. The process of plucking starts when meltwater produced near the bottom of the ice finds its way into cracks in the underlying rock. As the water refreezes and expands, it loosens chunks of rock, which are then torn away and carried by the glacier. Plucked material can range in size from silt to boulders the size of small apartment buildings.

Additional evidence for the combined power of plucking and

Figure 33. Whaleback on Rainy Lake, NPS.

abrasion can be found in strange polished oblong knobs that protrude above the surface of lakes and swamps. These look like small glacially sculpted hills and are called whalebacks (fig. 33). Whalebacks in the park are generally composed of light-colored granite, which is harder and more resistant to glacial scouring than biotite schist or volcanic rocks. They tend to have smooth, polished, and striated ramps pointing in the direction from which the ice came. On the opposite side they are rough and sometimes

steeper. Whalebacks form as ice encounters resistant knobs of bedrock. The force of the flowing ice on the front side of the knob abrades and polishes the rock—creating the smooth, gentle ramp you see on one end of the whaleback. The force of ice colliding with the rock also puts the ice under greater pressure, and this greater pressure may melt some of the ice. The water produced by this "pressure melting" finds its way into cracks in the rock on the side of the knob pointing away from the direction the ice came. The water then refreezes and loosens chunks of rock, which are easily plucked.

All the sediments that the glacier picks up by abrading and plucking are dropped when the glacier melts. One of the most common kinds of deposits is called glacial till. Glacial till consists of unsorted debris, silt, clay, sand, and rocks of different shapes and sizes. The material that makes up till is composed of a variety of minerals, each reflecting the different types of bedrock across which the ice moved. Glacial till drapes many low-lying areas of the park. Because vegetation covers most of the low-lying areas, the remnants of the glacial till are best exposed along the shorelines of the lakes. Over the years waves have washed away the fine clay and silt to produce spectacular sand beaches as seen in figure 34. Along particularly windy shorelines, erosion of the sand has left behind only cobbles.

Sometimes glaciers will transport very large boulders hundreds of miles. When the glacier drops the boulders, they may look out of place in their new surroundings because they are made of rock much different from the rock upon which they sit. These boulders are called glacial erratics. Glacial erratics occur throughout the park, usually in odd places, like the top of a ridge. A few of the more spectacular erratics in Voyageurs National Park can be seen in figures 35a, 35b, and 35c.

Facing Page: Figure 34. Sand beach, Voyageurs National Park, Raymond Klass © 2006.

A few other common glacial features are worth mentioning, even though they are found south of the park—the first being an end moraine. When a glacier melts, it will leave a pile of sediments at the farthest terminus. The spine of this pile of sediments will be

Above: Figure 35a. Glacial erratic, Oslo Lake, Kabetogama Peninsula, NPS.

Facing top: Figure 35b. Glacial erratic, Cranberry Bay, Rainy Lake, NPS.

Facing bottom: Figure 35c. Glacial erratic, Hoist Bay, Namakan Lake, NPS.

oriented perpendicular to the direction the glacier is moving. An end moraine close to the park is near Orr, Minnesota, approximately thirty miles south of Crane Lake. Visitors traveling north to Orr on Highway 53 will crest this moraine just outside of town. The moraine marks the terminal advance of a portion of the Rainy Lake Lobe that developed during the Wisconsin glaciation (we will learn more about the Rainy Lake Lobe in the next section).

A melting glacier produces an enormous amount of water. This water will wash sediments from the glacier into wide, braided stream channels where the sediments will be rounded and polished. Meltwater streams also sort glacial sediments by size: Larger, heavier pieces such as gravel will fall out before finer materials, like clay and silt. In Minnesota, the braided streams redeposited this sorted material into large flat beds called glacial outwash (fig. 36).

Glacial periods are not static; that is to say glaciers do not just form, expand, and then stay in place, like a snowbank. Instead, they tend to advance, then melt and recede, advance, then melt and recede. Snow that piles up on glaciers fuels the advances. Because the weather patterns responsible for dropping snow change over time, the geographic location where the most snow falls also changes over time. Where snow falls determines the direction of glacial advance. If snow is piling up to the northwest,

Figure 36. Outwash plain, Denali National Park, AK, USGS.

the advancing glacier will come from the northwest. If snow is piling up to the northeast, the glacier will come from that direction.

In the park, there were probably as many as six glacial advances during the Wisconsin glaciation, but only the records of the last two advances are preserved. Evidence for the other advances comes from other parts of the Midwest.

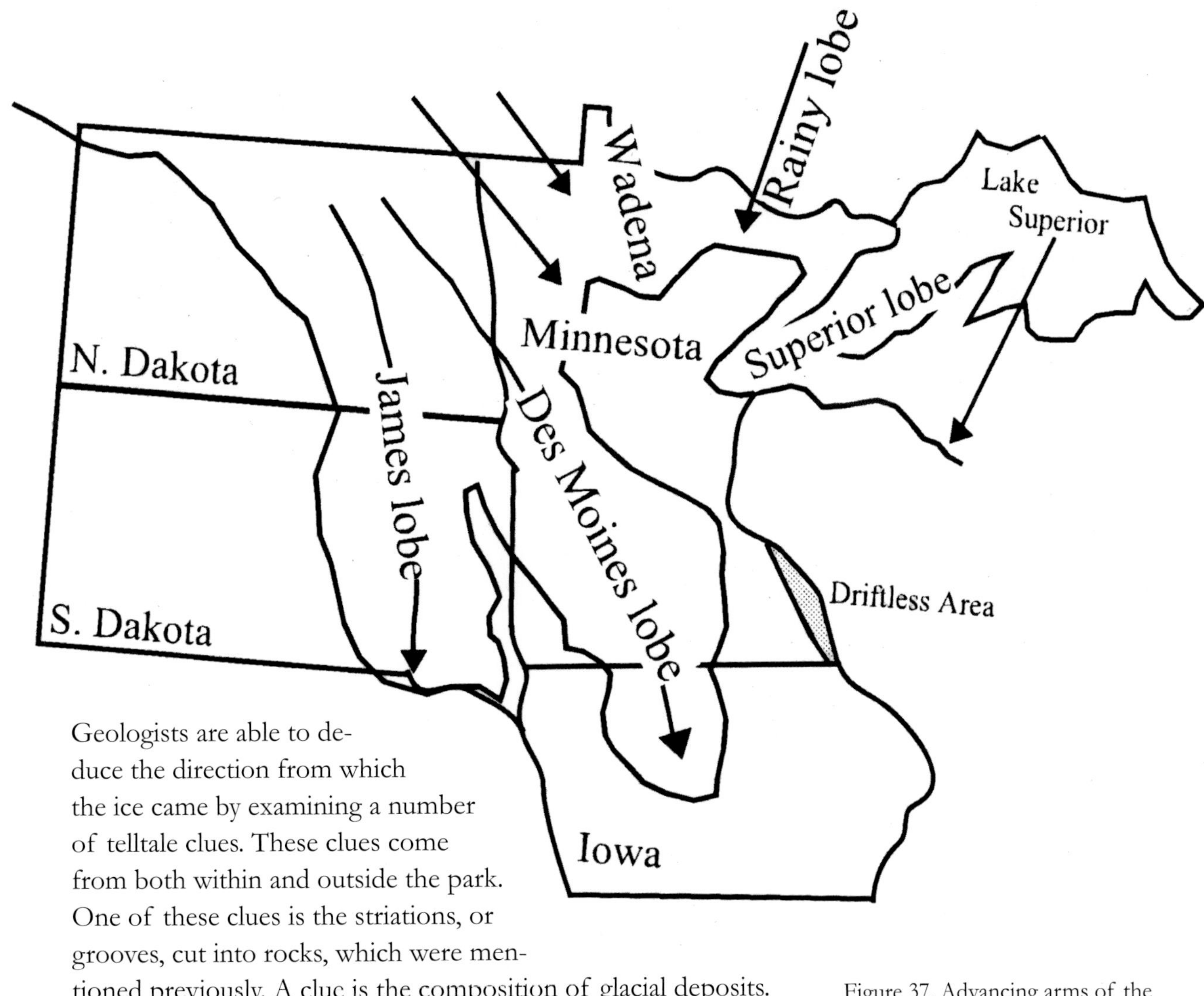

Geologists are able to deduce the direction from which the ice came by examining a number of telltale clues. These clues come from both within and outside the park. One of these clues is the striations, or grooves, cut into rocks, which were mentioned previously. A cluc is the composition of glacial deposits. Glaciers coming from different directions carry sediments from different parent rocks. Identifying where the parent rocks are found will tell geologists where the glacier passed before arriving in the park. Using this evidence, geologists have constructed the following history.

Figure 37. Advancing arms of the Laurentide Ice sheet. *Roadside Geology of Minnesota*, Richard. W. Ojakangas.

About 30,000 years ago, an arm of the Laurentide Continental Ice Sheet called the Rainy Lobe (advancing arms are called lobes) advanced across the park (fig. 37). Glacial striations found in polished bedrock in the central and eastern portions of the park are oriented to the northeast at about thirty degrees, parallel to the path of the ice lobe (fig. 32). The Rainy Lobe was probably separated from the Superior Lobe by highlands west of Lake Superior. The Rainy Lobe persisted until about 15,000 years ago. During this 15,000-year interval, a mile-thick sheet of ice swept the park. Great chunks of the softer rocks, like the biotite schists, were eroded.

Soon after the Rainy Lobe melted, the Laurentide Continental Ice Sheet made one final passage across the borders of the park. The Koochiching Lobe was sent down from the northwest. It ploughed across the western end of the park in a final burst of sluggish ice near the end of the Wisconsin time, thrusting as far to the east as the Kabetogama Lake Visitor Center. Whalebacks indicate its direction of passage. Stagnation of the ice, about 12,000 years ago, eventually produced deposits called Koochiching till, shown in relation to the Rainy Lobe till (fig. 38).

Visitors can see Koochiching till at the water's edge near Gold Portage, between Black Bay and Kabetogama Lake. The till is generally gray in color, but it contains buff to white limestone pebbles and cobbles. Geologists have traced this limestone back to its source in the Winnipeg lowland of Manitoba. Caliche is a type of accumulation associated with arid climates and is locally present in the Koochiching till. This is a good indication that at some time after the ice had melted, the climate of the region was warmer and more arid than at present.

Figure 38. Glacial till, NPS.

Glaciers are the great shapers of landscapes but their influence is ultimately subject to plate tectonics. The advancing lobes of the continental ice sheet and the erosive power of glacial meltwater carved most of the lakes in the park to their current size and shape. The location of these lakes is not accidental, but is a legacy of the ancient, 2.7-billion-year-old landscape. Glaciers flow like water, following low points in the landscape. Like water, glaciers also cut into the most easily eroded rock. The easiest bedrock to erode is rock already weakened by fractures, such as fault rock. This is true within the park, where most of the glacially scoured lakes are underlain by ancient faults. For example, the Rainy Lake–Seine River Fault Zone is mainly underwater. And a close look at the Geologic Map of Voyageurs will show that most of the smaller fault zones occupy linear areas of low topographic relief. Often these fault traces contain creeks, beaver ponds, and lakes. In Chapter 2, it was stated that the force of colliding plates about 2.7 billion years ago created the faults within the park. These faults provided the template upon which the glaciers could shape the surface of the park.

Chapter 12

FROM ICE TO WATER: THE PAST 12,000 YEARS

It is wonderful how abundant are those peculiar memorials of ice-action which neither the landslip, nor the avalanche, nor the vibrations of earthquake, nor any known agent but ice, can produce; and which the torrent, whether of mud or water cannot cause, but on the contrary immediately effaces.
Journal entry, September 10, 1857
Northern Italy, - Charles Lyell

If you visit Gold Portage to see the Koochiching till, continue a little farther until you reach the rapids. Close your eyes and listen to the roar of water coming through the rapids. Imagine that it is but a lingering echo of melting ice sheets. Imagine glacial meltwater cascading through narrow places of fault-driven rock, places that look similar to Gold Portage, cutting deeper and wider channels carved by glacial ice.

The last chapter discussed how water in its solid form—ice—shaped the landscape. As global temperatures rose, the Laurentide Continental Ice Sheet rapidly met its demise. The ice turned to water. The meltwater from the great glaciers eroded rock, moved huge deposits of clay and other sediments,

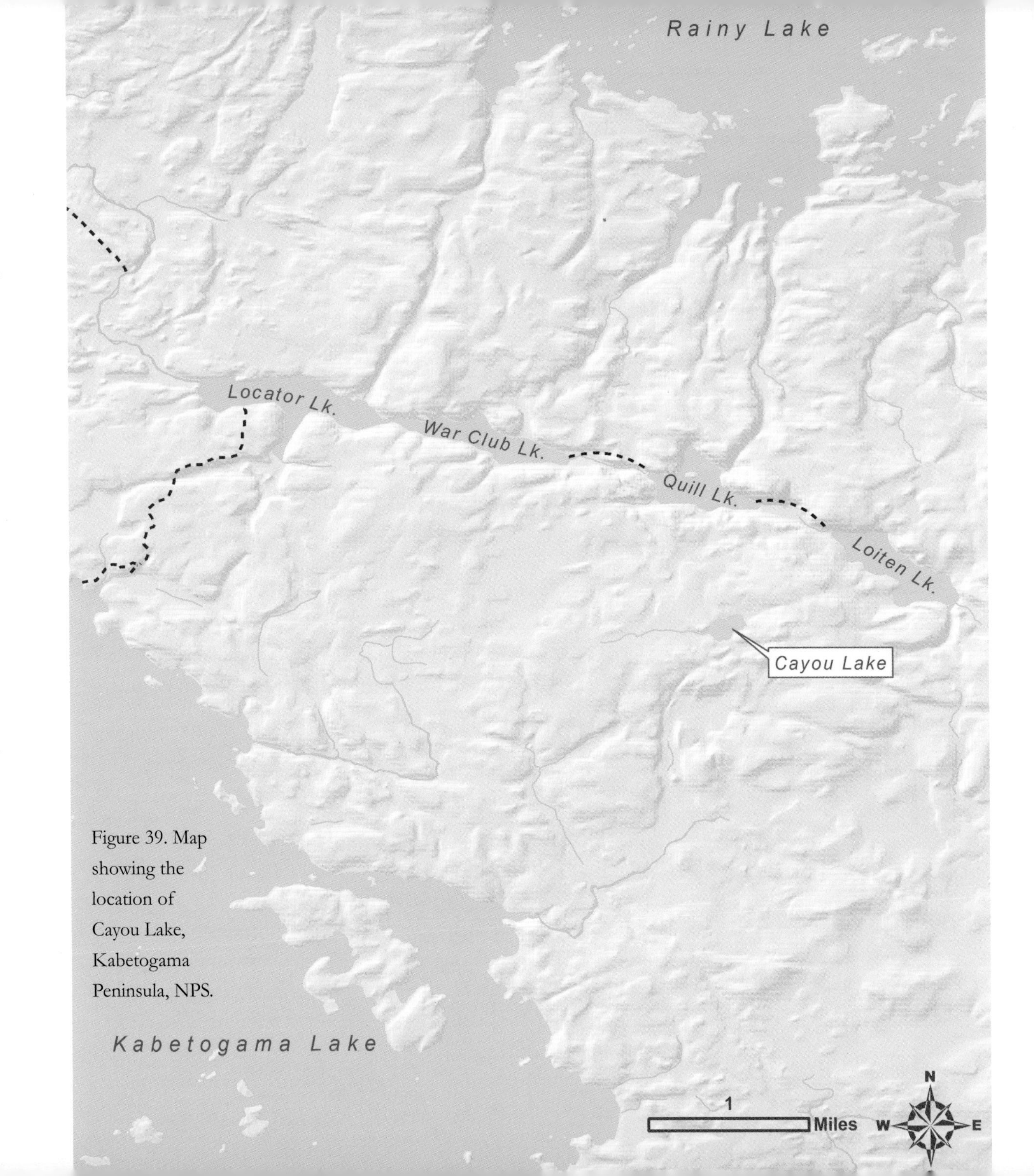

Figure 39. Map showing the location of Cayou Lake, Kabetogama Peninsula, NPS.

channeled the migration of plants and animals, and shaped the cultures of human beings. The water also created a record of all that it had accomplished.

Cayou Lake, a small tea-colored lake just to the south of Quill Lake on the Kabetogama Peninsula, has caught the attention of scientists (fig. 39). Although the scenery around the lake is beautiful to the eye, more attractive to scientists are layers of sediments found on the lake's bottom. These sediments record all of the postglacial and part of the glacial history of the park.

Facing Page: Figure 40. Glacial Lake Agassiz, *Roadside Geology of Minnesota*, Richard W. Ojakangas.

Under the right conditions, sediments pile up on the bottom of a lake in layers like a bundle of roofing shingles. This event is rare. Most times, an agent, such as wind, waves, flood, or drought, mixes up the layers and makes the record messy. Cayou Lake is a place where layers from years past have neatly piled one on top of the other. Two scientists, M. G. Winkler and P. R. Sanford, collected cores of bottom sediments from Cayou Lake and attempted to tease out and order clues about the return of northern trees, the expansion of glacial lakes, the frequency of forest fires, and the arrival of European farmers.

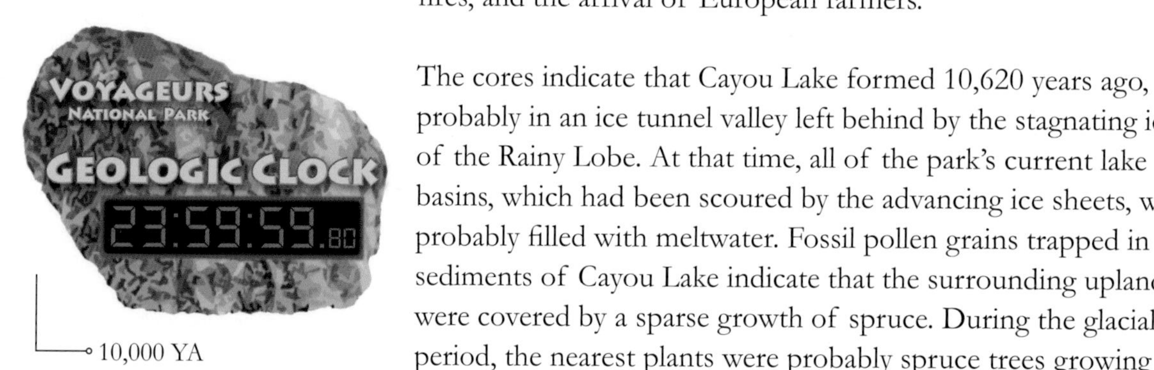

The cores indicate that Cayou Lake formed 10,620 years ago, probably in an ice tunnel valley left behind by the stagnating ice of the Rainy Lobe. At that time, all of the park's current lake basins, which had been scoured by the advancing ice sheets, were probably filled with meltwater. Fossil pollen grains trapped in the sediments of Cayou Lake indicate that the surrounding uplands were covered by a sparse growth of spruce. During the glacial period, the nearest plants were probably spruce trees growing

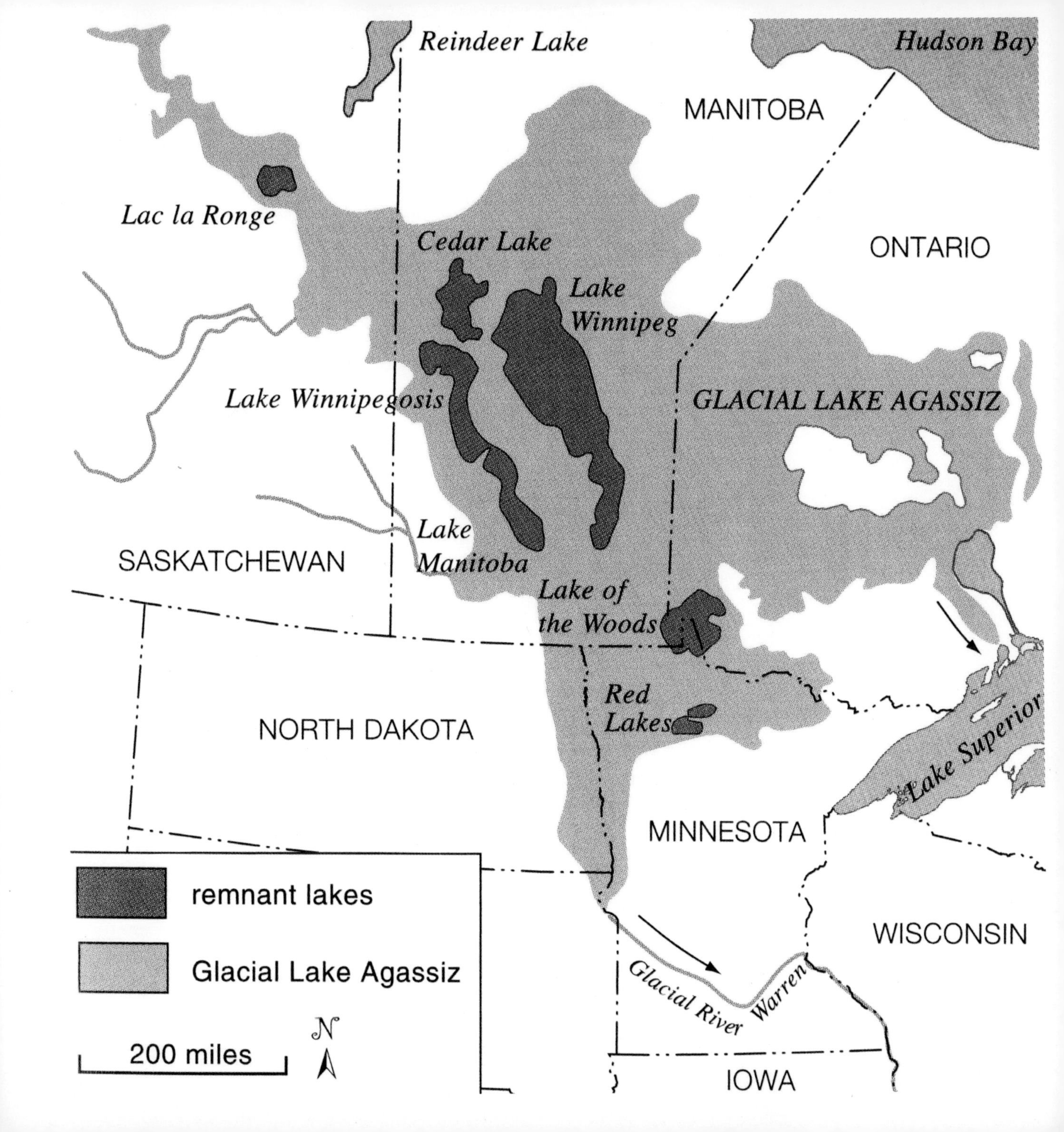

Reindeer Lake
Hudson Bay
MANITOBA
Lac la Ronge
Cedar Lake
ONTARIO
Lake
Winnipeg
Lake Winnipegosis
GLACIAL LAKE AGASSIZ
Lake
Manitoba
SASKATCHEWAN
Lake of
the Woods
Red
Lakes
NORTH DAKOTA
Lake Superior
MINNESOTA
remnant lakes
Glacial Lake Agassiz
WISCONSIN
Glacial River Warren
N
200 miles
IOWA

on the southern margin of the glaciers. After the glaciers melted, they were the first trees to recolonize the landscape.

Glaciers often serve as dams and block the drainage of their own meltwater. The lakes that can fill up behind these dams are called glacial lakes and can be more than a hundred meters deep. According to Winkler and Sanford, the rising waters of Glacial Lake Agassiz, the largest glacial lake to occupy the North American continent, inundated the park about 9,900 years ago (fig. 40).

Glacial Lake Agassiz itself formed between about 12,000 and 11,000 years ago, when the Des Moines Lobe melted back into the Red River Lowland. The Red River Basin is a north-flowing drainage, meaning that draining meltwater was trapped by the retreating ice margin. To the south, a topographic divide near Browns Valley in western Minnesota formed another dam. Rising meltwater quickly filled the lowlands until it finally overtopped the southern dam, creating an outlet that is known as the River Warren. The southeast flowing torrents rapidly carved what is now the Minnesota River Valley. Agassiz drained through the River Warren until about 9,200 years ago. About that time the receding ice opened up another outlet through the Lake Superior basin and drained the lake to a level low enough to leave the River Warren dry.

At the time Agassiz covered portions of the park, it was at its maximum size, about 135,000 square miles, and stretched across Ontario, Manitoba, Minnesota, and the Dakotas. Fish and other aquatic animals used Agassiz to spread throughout the recently deglaciated landscape. The distribution of fish and many of the aquatic animals found in the lakes of Voyageurs, and indeed, much of northern Minnesota, Ontario, and Manitoba, are directly related to flooding by Glacial Lake Agassiz. Fossils of microscopic organisms found in Cayou Lake, like diatoms and water fleas, confirm this.

About 9,000 years ago, Agassiz had drained enough to uncover Cayou Lake and probably many other areas of the park. Spruce forest returned after the lake left, but so did two species of trees that had spent the glacial period farther south, jack pine and red pine. This is a fact that visitors should stop to consider. When traveling the waters of Voyageurs, it is easy to look upon the shoreline forests of pine and think of them as primordial. In fact, the red and jack pines have only been back here for about 9,000 years, and probably arrived after we humans did. They, along with many other plants that populate the park, spent the Ice Age far to the south. Their seeds, carried by animals, wind, and water, slowly moved northward to recolonize the postglacial landscape.

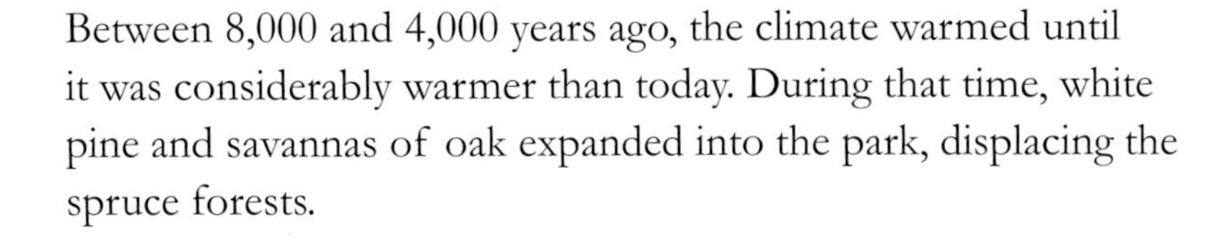

Between 8,000 and 4,000 years ago, the climate warmed until it was considerably warmer than today. During that time, white pine and savannas of oak expanded into the park, displacing the spruce forests.

Beginning about 4,000 years ago, the climate changed again. This time it became cooler and wetter. Spruce, fir, and tamarack moved in. The savannas of oak shrank, but the pines hung on to their place in the forests. Sphagnum bogs began to expand in the wet, cool climate. One such bog that lies partially within the southwestern corner of the park is the East Rat Root River Peat-

10,000 YA

land. The State of Minnesota has designated 2,732 acres of this peatland a Scientific Natural Area.

One of the more intriguing remains found in sediments from Cayou Lake is charcoal. Charcoal, the remains of scorched plants, provides a record of past fires. About 4,000 years ago, charcoal increased in the sediments, indicating an uptick in the number and size of fires. This is puzzling because the climate was cooler and wetter. Some researchers have suggested an answer to this seeming contradiction. They believe that 4,000 years ago, human inhabitants of the region may have started using fire as a tool to manage the forests. Fires encourage the growth of edible berries. Fires also spark the sprouting and growth of saplings, food for moose, deer, and other grazing animals that people hunted.

There is other evidence that people were thriving in North America immediately following the retreat of the glaciers. This evidence comes not from lake sediments but from the mammalian fossil record.

At the end of the Ice Age, many species of animals probably migrated north along the margins of the retreating ice sheets and entered what is now Voyageurs National Park. Many of those animals—wolves, beavers, black bears, deer, and moose—still inhabit the park today. Others are extinct. The extinct animals include mastodons, woolly mammoths, saber-tooth cats, dire wolves, giant beavers, woodland musk oxen, short-faced bears, and many smaller animals. In all, about seventy species went extinct in North America at the end of the Ice Age. Glacial scraping and later flooding and erosion have likely destroyed or buried most fossil evidence within the park and most of Minnesota, but the few fossil remains in the northern states indicate that many of these extinct mammals inhabited latitudes that include the park. Where did they go? What happened to them?

Most of the seventy species mentioned above were extinct by about 11,000 years ago. People are believed to have arrived in North America no later than about 14,000 years ago. One hypothesis is that the arrival of human hunters in a land where none were previously known led to the extinction of a large number of mammals. This hypothesis is called the overkill hypothesis. Although current evidence makes it one of the stronger explanations for the extinctions, not all scientists adhere to it. Other scientists have proposed that rapid climate change at the end of the Ice Age or widespread disease killed the large mammals. While there is less evidence to support these alternative hypotheses, they very well may have acted in concert with human hunting to bring an end to many mammal species.

Black bear in Voyageurs National Park, NPS.

Reading the postglacial history of humans, plants, and animals in the region of the park should give us a great appreciation for how much our planet changes and how impermanent the forms of life we currently share the planet with are. The plants and animals that now occupy the park reinvaded no more than 12,000 years ago. During that time, lakes expanded and contracted, rivers changed their courses, and a giant continental glacier melted and drained into the sea. Seventy species of mammals, some of them the largest to occupy the continent, went extinct. People witnessed all of this.

Since the retreat of the glaciers, humans have shaped and been shaped by the landscape. Migrating cultures used the waterways as transportation routes. Many of the postglacial plants and animals of the region—wild rice, moose, deer, sturgeon, caribou, and berries, to name a few—provided food for the people who came to live here. People likely managed the forests with large fires aimed at encouraging the growth of berries or the proliferation of wildlife. When Europeans arrived 400 years ago, what greeted them was not the primeval, unpeopled landscape of myth, but a thriving Native American culture that had been managing and occupying the lakes and forests for more than 9,000 years. In the brief time since Europeans came, they have wrought their own form of change on the land. They cleared forests and then watched some of them recover. They hunted a few animals to local extinction. They dug gold out of the ground and built cities in anticipation of the wealth that would follow.

The shape of everything people have seen over the past 11,000 years has its roots ultimately in the rocks of the park. The geologic history of the park determined where lakes formed, where gold was found, and where trees would grow.

And so we find Voyageurs today: a place originating from the fire and heat of a primordial planet. The current crust of the park was carried skyward on the smoky peaks of volcanoes. It was carried across the surface of the planet, piggyback on a small continental plate. It collided with other island arcs and pushed up lofty mountains. These mountains towered over the Archean landscape, only to be brought low by the slow relentless pressure of weathering and erosion. Colliding plates tore apart the earth and folded it into a series of corrugations like those on a cardboard box.

The geologic tale of Voyageurs is sometimes violent, sometimes silent, but never ends. The ice will likely reinvade and the Canadian Shield will likely rift apart. The overall plot is change, not for good or ill, just change, daily, yearly, through billions of years. Every rock that we pick up is a window into the past. Our knowledge of geologic processes allows us to wipe away the encrusted dust and grime to peer through the pane of time into a turbulent world lit by the bright light of our imagination.

Deer in Voyageurs National Park, Teri Tucker.

GLOSSARY

Accretionary Wedge	A mountain range formed as sedimentary strata and hard crust are scraped off the top of subducting plate.
Alluvial Fan	Sedimentary material eroded from a steep mountain front, carried through a canyon and deposited by flash floods in a fan-shaped pattern onto the flat land beyond the canyon mouth.
Basalt	A fine-grained, dark-colored igneous rock with about 50 percent silica.
Basaltic Greenstone	In greenstone, the olivine and peridotite that made up the fresh basalt have been metamorphosed by high pressure and warm fluids into green minerals—epidote, actinolite, or chlorite, depending on the exact conditions. The white mineral is aragonite, an alternative crystal form of calcium carbonate (its other form is calcite).
Biotite Schist	Any of various medium-grained to coarse-grained metamorphic rocks composed of laminated, often flaky parallel layers of chiefly micaceous minerals.
Boreal Forest	An evergreen coniferous forest of northern North America dominated by fir, pines, and spruces.
Canadian Shield	A plateau region of eastern Canada extending from the Great Lakes and the St. Lawrence River northward to the Arctic Ocean. The highland formation also covers much of Greenland and forms the Adirondack Mountains in the United States.
Compression	The decrease in volume of a material as it is squeezed.
Contact Metamorphism	The process by which rocks are altered in composition, texture, or internal structure by extreme heat, pressure, and the introduction of new chemical substances.
Continental Crust	Outer layer of the Earth that is thick and high in silica compared to oceanic crust.

Convection Currents	Mass movement of subcrustal or mantle material as the result of temperature variations.
Craton	Relatively flat-lying region of a continent that has not experienced widespread tectonic activity for hundreds of millions of years.
Dacite	A fine-grained igneous rock with about 65 percent silica (more silica than andesite, but less than rhyolite).
Dike	A sheet-like, ingenious rock body formed where magma cuts across rock layers.
Eon	A major portion of geologic time, broken into smaller divisions called eras.
Epoch	A division of a period of geologic time.
Era	A division of an eon of geologic time, broken into smaller units called periods.
Evolution	Change in the genetic composition of a population during successive generations, as a result of natural selection acting on the genetic variation among individuals, and that could result in the development of new species.
Fault Zone	A fault expressed as an area of numerous small fractures. Also known as distributed fault.
Geochronologist	The chronology of the Earth's history as determined by geologic events.
Glacier	A naturally formed mass of ice that moves gradually down slope under the influence of gravity.
Gneiss	A metamorphic rock formed form pre-existing rock that underwent extreme increases in temperature and pressure. You can see how the mineral grains in the rock were flattened through tremendous heat and pressure and are arranged in alternating patterns.
Granite Sill	A tubular pluton that intrudes between older layers of sedimentary rock. Sills are parallel to beds of surrounding rock, usually horizontal in nature.

Greenstone Belt	Are zones of variably metamorphosed mafic to ultramafic volcanic sequences with associated sedimentary rocks that occur within Archaean and Proterozoic cratons between granite and gneiss bodies.
Greenstone	A type of metamorphic rock rich in green tinged minerals. Commonly, old (Precambrian) rocks formed from the metamorphism of ancient oceanic crust.
Island Arc	A curved chain of volcanoes formed where a tectonic plate capped by oceanic crust subducts beneath another plate capped by oceanic crust.
Kenoran Orogeny	A Precambrian thermal event on the Canadian Shield that occurred 2.5 billion years ago (± 150 million years). Rocks affected by the Kenoran event represent some of the oldest rocks in North America and occur in the Superior Province surrounding Hudson Bay on the south and east, the Slave Province in northwestern Canada, and the small Eastern Nain Province on the northeastern Labrador coast. Beyond the shield area, rocks equivalent in age to Kenoran rocks occur in Wyoming and the Black Hills. Parts of the Lewisian Gneiss of Scotland have been dated at 2.4 billion to 2.6 billion years ago, and these have been correlated with the Kenoran thermal event.
Lapilli	Size classification term for tephra, which is material that falls out of the air during a volcanic eruption or during some meteorite impacts.
Mafic Tuff	Mafic tuff horizons were deposited and resedimented in mostly terrestrial settings. Initial pyroclastic volcanism is attributed to interaction between rising mafic magma batches and shallow ground water or surface water. Overlying flood basalt piles represent relatively rapid outpourings of mafic lava that complete the volcanic cycle.
Magma	Hot, liquid rock that may contain some gas and solid material.
Magma Chamber	An accumulation of molten rock below Earth's surface.
Melange	A chaotic mixture of broken and jumbled rock, commonly formed within the accretionary wedge at a subduction zone.
Metamorphism	Alteration in the structure and/or composition of rocks due to pressure and/or temperature.

Metasedimentary	Sedimentary rocks altered by metamorphism.
Metasedimentary Belt	It comprises turbidites and continental sandstones, both of which are interlayered with rare volcanic rocks and intruded by plutons.
Mid-Ocean Ridges	An undersea mountain range formed from volcanic activity where lithospheric plates diverge. The mid-ocean ridge is an interconnected system of undersea volcanoes that meander over the Earth like the raised seams on a baseball. It is a continuous 40,000-mile (60,000-kilometer) seam that encircles Earth and bisects its oceans. The mid-ocean ridge represents an area where, in accordance with plate tectonic theory, lithospheric plates (also called tectonic plates) move apart and new crust is created by magma (molten rock) pushing up from the mantle. The mid-ocean ridge system is an example of a divergent (rather than a convergent or transform) plate boundary.
Mid-Ocean Rift	General term for an underwater mountain system that consists of various mountain ranges (chains), typically having a valley known as a rift running along its spine, formed by plate tectonics. This type of oceanic ridge is characteristic of what is known as an oceanic spreading center, which is responsible for sea floor spreading.
Migmatite	Migmatites form under extreme temperature conditions during prograde metamorphism, where partial melting occurs in pre-existing rocks. Migmatites are not crystallized from a totally molten material, and are not generally the result of solid-state reactions. Migmatites are composed of a leucosome, new material crystallized from incipient melting, and a mesosome, old material that resisted melting. Commonly, migmatites occur within extremely deformed rocks that represent the base of eroded mountain chains, typically within Precambrian cratonic blocks.
Oceanic Crust	Outer layer of the Earth that is thin and low in silica compared to continental crust.
Orogeny	A geological event that forms mountains.
Period	A division of an era of geologic time, broken into smaller units called epochs.
Pikwitonei Subprovince	Granulite rich crust (Pikwitonei) forms the northwestern boundary (Nelson Front) of Superior Province and underlies rare small interior domains.

Pillow Lava — A rock layer (lava flow) with globular structures formed when magma erupted through the seafloor or flowed in the ocean.

Plate Tectonics — The theory that features on Earth's surface result from the horizontal movements of large plates of Earth's outer shell.

Platform — The part of the continental craton that is periodically invaded by shallow seas and covered by sedimentary deposits.

Playa Lake — Basins of internal drainage. In arid regions with basins of internal drainage rainfall in the adjacent areas is carried into the basin by ephemeral streams carrying water and dissolved ions. The water fills the low points in the basin to form a playa lake. These lakes eventually evaporate, resulting in the precipitation of salts such as halite, gypsum, anhydrite, and a variety of other salts not commonly found in marine evaporate deposits.

Conglomerate — Interbedded polymictic conglomerates represent interruption in platform evolution during onset of the Klakas orogeny, an arc-continent collisional event that occurred in the Late Silurian–Early Devonian periods.

Province — Refers to a specific physiogeographic area that comprises a grouping of like bathymetric or former bathymetric elements (now sedimentary strata above water) whose features are in obvious contrast to the surrounding regions, or other "provinces." The term usually refers to sections or regions of a craton recognized within a given time-stratigraphy, i.e., recognized within a major division of time within a geologic period.

Rhyolite — A fine-grained, generally light-colored igneous rock with about 70 percent silica. (It is the extrusive equivalent of granite.)

Schist — A metamorphic rock formed from a pre-existing rock that experienced relatively large increases in temperature and pressure. (Schist represents a greater amount of metamorphism than slate, but less than gneiss.)

Subduction Zone — A convergent plate boundary occurs where one plate slides deeply beneath another.

Turbidites	A sequence of thick sandstone and thin shale layers deposited by a turbidity flow.
Volcanic Ash	Fine-grained pyroclastic material, blown from a volcano and carried away and deposited by winds.
Volcanic Bomb	A large piece of pyroclastic material, commonly football- to watermelon-size, that may have developed the shape of a bomb as it traveled through the air in a liquid state.
Volcanic Island Arcs	Formed from the subduction and melting of oceanic crust as it descends into the mantle underneath a less dense oceanic crust at a convergent plate boundary. The subduction results in the creation of undersea volcanoes which then rise above sea level. The resulting volcanoes create a string of islands called an island arc. The curve of an island arc echoes the curve of its deep-ocean trench.
Whaleback	Streamlined bedrock landforms that result from glacial quarrying and abrasion.

REFERENCES

Alminas, H.V., Foord, E.E., Cartwright, D.F., and Dahlberg E.H., "Geochemical Studies in the Vermilion District and the Duluth Complex, Northeastern Minnesota—A Progress Report," *Institute on Lake Superior Geology* 39, part 1: 4-5.

Anhaeusser, C.R., Mason, Robert, Viljoen, M.J., and Viljoen, R.P., "A Reappraisal of Some Aspects of Precambrian Shield Geology," *Geol. Soc. Am. Bull.* 80, no. 11, 1969: 2175-2200.

Ashwal, L.D., Wooden, J.L., Phinney, W.C., and Morrison, D.A., "Sm-Nd and Rb-Sr Isotope Systematics of an Archean Anorthosite and Related Rocks from the Superior Province of the Canadian Shield," *Earth and Planetary Science Letters* 74, 1985: 338-346.

Ault, David, "A Review of the Geology of Rainy Lake" (Minneapolis: Univ. of Minn. M.S. Thesis, 1958): 73.

Balaban, N.H. (Ed.), "Field Trip and Guidebook for Selected Areas in Precambrian Geology of Northeastern Minnesota, Geological Soc. Am. North-Central Section 21st Annual Meeting" (St. Paul: Minn. Geol. Surv., Guidebook Series, No. 17): 1987.

Barnett, P.J., "Quaternary Geology of Ontario," *in* Thurston, P.C., Williams, H.R., Sutcliffe, R.H., and Stott, G.M., eds., *Geology of Ontario* (Ontario Geological Surv., Special Volume 4, Part 2, 1992), 1,011-1,088.

Bauer, Robert L., "The Petrology and Structural Geology of the Western Lake Vermilion Area" (Minneapolis: Univ. of Minnesota, Northeastern Minnesota, Ph.D. Thesis, 1981).

Bauer, Robert L., "Correlation of Early Recumbent and Younger Upright Folding Across the Boundary Between Archean Gneiss Belt and Greenstone Terrane, Northeastern Minnesota," *Geology* 13 (1985): 657-660.

Bauer, Robert L., "Multiple Folding and Pluton Emplacement in Archean Migmatites of the Southern Vermillion Granitic Complex, Northeastern Minnesota," *Can. Jour. Earth Sci.* 23 (1986): 1553-1764.

Bauer, Robert L., "Contrasts in the Development of Concomitant Structures Along the Faulted Boundary Between an Archean Gneiss Belt and Greenstone Belt, NE Minnesota," *Geol. Soc. Am. Abstracts with Programs* 19 (1987): 188.

Bauer, Robert L., "Multiphase Folding During Batholith Emplacement in a Single Shortening Event, Vermilion Granitic Complex, NE Minnesota, Geological Association of Canada—Mineralogical Association of Canada," *Program with Abstracts* 13 (1988): A6.

Bauer, Robert L., and Bidwell, M.E., "Contrasts in the Response to Dextral Transpression Across the Quetico-Wawa Subprovince Boundary in Northeastern Minnesota," *Can. Jour. Sci.* 27 (1990): 1521-1535.

Bauer, Robert L., and Hudleston, Peter J., and Southwick, David L., "Structural Evolution of the SW Quetico Belt in Northern Minnesota, Geological Association of Canada—Mineralogical Association of Canada," *Program with Abstracts* 14 (1989): A55.

Bauer, Robert L., and Hudleston, Peter J., and Southwick, David L., "Deformation Across the Western Quetico Subprovince and Adjacent Boundary Regions in Minnesota," *Canadian Jour. of Earth Sciences* 29 (1992): 2,087-2,103.

Beck, Warren, and Murphy, V.R., "Rb-Sr and SM-Nd Studies of Proterozoic Mafic Dikes in Northeastern Minnesota [abs.]," *Proceedings*, 28th Institute on Lake Superior Geology, International Falls, Minn., 1982, 5.

Biek, Robert, "Warped Beaches of Glacial Lake Agassiz," *in* Biek, R., ed., *North Dakota Geol. Surv. Newsletter* 20, no. 3 (1993): 7-12.

Blackburn, C.E., Johns, G.W., Ayer, J., and Davis, D.W., "Wabigoon Subprovince," *in* Thurston, P.C., Williams, H.R., Sutcliffe, R.H., and Stott, G.M., eds., *Geology of Ontario, Ontario Geological Surv.*, special volume 4, part 1 (1991), 303-381.

Boerboom, T.J., and Zartman, R.E., "Geology, Geochemistry, and Geochronology of the Central Giants Range Batholith, Northeastern Minnesota," *Canadian Jour. of Earth Sciences* 30, no. 12 (Dec. 1993): 2510-2522.

Canada Department of Environment, Rainy Lake (Northern) Redgut Bay, Can. Dept. Environ., Marine Sciences Branch, Hydrologic Service, Chart 6110, scale 1:25,000, 1968b.

Canada Department of Environment, Rainy Lake (Eastern) Seine Bay to Sturgeon Falls, Can. Dept. Environ., Marine Sciences Branch, Hydrologic Service, Chart 6111, scale 1:25,000, 1968c.

Canada Department of Environment, Rainy Lake (Western) Fort Frances to Hostess Island and Sandpoint, Can. Dept. Environ., Marine Sciences Branch, Hydrologic Service, Chart 6108, scale 1:25,000, 1970a.

Card, K.D, "A Review of the Superior Province of the Canadian Shield, a Product of Archean Accretion," *Precambrian Research* 48 (1990): 99-156.

Chandler, V.W., and Horton, R.J., "Complete Bouguer Gravity Anomaly Map of the International Falls 1°x2° Quadrangle, Minnesota and Ontario," *U.S. Geol. Surv. Open-File Rept.* (1988): 88-265.

Craddock, J.P., Kennedy, B.C., Cook, A.L., Pawlisch, M.S., Johnston, S.T., Jackson, M., "Anisotropy of Magnetic Susceptibility Studies in Tertiary Ride-Parallel Dykes (Iceland), Tertiary Margin-Normal Aishihik Dykes (Yukon), and Proterozoic Kenora-Kabetogama Composite Dykes (Minnesota and Ontario)," *Tectonophysics* 448 (2008): 115-124.

Davidson, Donald M., Jr., "Geological Evidence Relating to Interpretation of the Lake Superior Basin Structure," *Geol. Soc. Am. Memoir* 156 (1982): 5-14.

Davies, J.C., Fort Frances Area, District of Rainy River, Ontario Dept. Mines, Prelim. Geol. Map P286, scale 1"=1 mile, 1965.

Davis, J.C., Geology of the Fort Frances Area, District of Rainy River, Ontario Dept. Mines, OFR 5013 (typescript), 41p. with Prelim. Geol. Map P286, scale 1"=1 mile, 1967.

Davies, J.C., and Pryslak, A.P., Kenora-Fort Francis Sheet, Geological Compilation Series, Ontario Dept. of Mines, 1967.

Davis, D.W., Poulsen, K.H., and Kamo, S.L., "New Insights into Archean Crustal Development from Geochronology in the Rainy Lake Area, Superior Province," *Canada, Jour. Geology* 97 (1989): 379-398.

Davis, J.C., Geology of the Fort Francis area, district of Rainy River, Ontario Division of Mines Geological Report 107, 35p (Map 2263, scale 1:63,360 [1"= 1 mi]), 1973.

Davis, Steven R., Hite, Alan G., and Larson, William S., Mineral occurrences and development potential near Voyageurs National Park, Minnesota, U.S. Bureau of Mines Mineral Land Assessment (MLA) Open File Report 5-94, 1994, 153p.

Davis, M., Douglas, C., Calcote, R., Cole, K.L., Winkler, M.G., Flakne, R., "Holocene Climate in the Western Great Lakes National Parks and Lakeshores: Implications for Future Climate Change," *Conservation Biology* 14, no. 4 (August 2000): 968-983.

Day, Warren C., Bedrock geologic map of the Rainy Lake area, northern Minnesota, U.S. Geol. Surv. Open-File Rept. 85-0246, 1 sheet, scale 1:24,000, 1985.

Day, Warren C., Bedrock geologic map of the Rainy Lake area, northern Minnesota, U.S. Geol. Surv. Misc. Investigations Map Series MI-1927, scale 1:50,000, 1987.

Day, Warren C., and Sims, Paul K., "Tectonic Evolution of the Rainy Lake Area, Northeastern Minnesota, Geological Association of Canada—Mineralogical Association of Canada," *Program with Abstracts* 9 (1984): 57.

Day, Warren C., Southwick, David L., Schulz, Klaus J., and Klein, Terry L., "Bedrock Geologic Map of the International Falls 1° x 2° Quadrangle, Minnesota, United States and Ontario, Canada," U.S. Geol. Surv. Misc. Investigations Series Map I-1965-B (1 sheet), 1990.

Drache, H.M., *Koochiching: The Rise and Fall of Rainy Lake City* (Danville, Ill.: Interstate Printers and Publication, Inc., 1983), 3-71.

Elson, John A., *Geology of Lake Agassiz, in Life, Land and Water—Proceedings of the 1966 Conf. on Environmental Studies of the Glacial Lake Agassiz Region, Univ. of Manitoba, 1966* (Winnipeg: Univ. of Manitoba Press, 1967), 36-96.

Englebert, J.A., Hauk, S.A., Southwick, David L., and Welsh, J.L., "Bedrock Geochemistry of Archean Rocks in Northern Minnesota," Natural Resources Research Institute, Technical Rept. 91/12, 1991, 200.

Fletcher, G.L., and Irvine T.N., "Geology of the Emo Area," Ontario Dept. of Mines Annual Report 63, 36p. (map 1954-2, scale 1:63,360 [1"=1mi]), 1954.

Fowler, John H., and Kuenzi, W. David, "Keweenawan Turbidites in Michigan (Deep Borehole Red Beds): A Foundered Basin Sequence Developed During Evolution of a Protoceanic Rift System," *Jour. Geophys. Res.* 83, no. B12 (Dec. 10, 1978): 5833-5843.

Frantes, J.R., "Petrology and Sedimentation of the Archean Seine Group Conglomerate and Sandstone, Western Wabigoon Belt, Northern Minnesota and Western Ontario (M.S. Thesis, Univ. of Minnesota-Duluth, 1987).

Fritz, David L., "Special History Study, Gold Mining Near Rainy Lake City from 1983 to 1901, A Theme Associated with Voyageurs National Park, Minnesota" (Denver: National Park Service, 1986), 127.

Frye, J.K., "Petrography of the Granites of the Minnesota-Ontario Boundary Region" (M.S. Thesis, Univ. of Minnesota, Minneapolis, 1959).

Goldich, S.S., "Geochronology of Minnesota," *in Geology of Minnesota: A Centennial Volume* (St. Paul: Univ. of Minn., Geol. Surv., 1972), 27-37.

Goldich, S.S., Nier, A.O., Baadsgaard, H., Hoffman, J.H., and Kruegar, E.W., "The Precambrian Geology and Geochronology of Minnesota," *Minn. Geol. Surv. Bull.* 41 (1961): 193.

Grout, F.F., "The Coutchiching Problem," *Geol. Soc. Am. Bull.* 36, no. 2 (1925): 351-364.

Grout, F.F., Petrographic Study of Gold Prospects of Minnesota," *Econ, Geol.* 32, no. 59 (1937): 56-68.

Grout, F.F., Gruner, J.W., Schwartz, G.M., and Thiel, G.A., "Precambrian Stratigraphy of Minnesota," *Geol. Soc. Am. Bull.* 62, no. 9, (1951): 1017-1078.

Halls, H.C., "Crustal Thickness in the Lake Superior Region," *Geol. Soc. Am. Memoir* 156 (1982): 239-243.

Hansen, G.N., "K-AR Ages from Granites and Gneisses and for Basaltic Intrusives in Minnesota," *Minnesota Geol. Surv., Rept. of Investigations,* no. 8 (1968): 20.

Harris, F.R., "Geology of the Rainy Lake Area, District of Rainy River," *Ontario Division of Mines, Geol. Rept.* 115 (1974): 94. (Maps 2278 and 2279, scale 1:31,680 [1"=½ mi]).

Hart, S.R., and Davis, G.L., "Zircon U-Pb and Whole-rock Rb-Sr Ages and Early Crustal Development Near Rainy Lake, Ontario," *Geol. Soc. Am. Bull.* 80, no. 4 (1969): 595-616.

Hinze, William J., and Wold, Richard J., "Lake Superior Geology and Tectonics—Overview and Major Unsolved Problems," *Geol. Soc. Am. Memoir* 156 (1982): 273-280.

Hooper, P., and Ojakangas, R.W., "Multiple Deformation in the Vermilion District, Minnesota," *Canadian Journal of Earth Sciences* 8 (1971): 423-434.

Horton, R.J., "Gravity Survey Data of the International Falls CUSMAP Area," *U.S. Geol. Surv. Open- File Rept. 87-0327* (1987): 52.

Horton, R.J., Day, W.C., and Bracken, R.E., "Aeromagnetic Survey of North-Central Minnesota," *in* Scott, R.W., Detra, P.S., and Berger, R.E., eds., *U.S. Geol. Surv. Bull.* 2039 (1993): 201-204.

Hudleston, Paul J., "Early Deformational History of Archean Rocks in the Vermilion District, Northeastern Minnesota," *Can. Jour. Earth Sci.* 13 (1976): 679-692.

Huddleston, Peter J., Schultz-Ela, D., Bauer, R.L., and Southwick, D.L., "Transpression as the Main Deformational Event in Archean Greenstone Belt, Northeastern Minnesota," *in Workshop on the Tectonic Evolution of Greenstone Belts, Lunar and Planetary Institute, Houston, TX, Technical Report 86-10* (1986): 62-63.

Huddleston, Peter J., Schultz-Ela, D., and Southwick, D.L., "Transpression in an Archean Greenstone Belt, Northern Minnesota," *Can. Journal of Earth Sciences*, 25 (1988): 1060-1068.

Jirsa, M.A., Southwick, D.L., and Boerboom, T.J., "Structural Evolution of Archean Rocks in the Western Wawa Subprovince, Minnesota: Refolding of Precleavage Nappes During D_2 Transpression," *Canadian Jour. of Earth Sciences* 29, no. 10 (1992): 2,146-2,155.

Johnston, W.A., "Rainy River District, Ontario, Surficial Geology and Soils," *Geol. Surv. Canada, Mem.* 82 (1915): 123.

Johnston, W.A., "The Genesis of Lake Agassiz: A confirmation," *Jour. of Geology* 24, no. 7 (1916): 625-638.

Kalliokoski, J., "Jacobsville Sandstone," *Geol. Soc. Am. Memoir* 156 (1982): 147-155.

Klein, T.L., Mineral Occurrence and Drill-Hole Location Map of the International Falls 1° x 2° Quadrangle, Minnesota and Ontario, U.S. Geol. Surv., Misc. Field Studies Map MF-2082, 1989.

Klein, T.L., *et al.*, "Geochemical Data from the International Falls and Roseau, Minnesota CUSMAP Projects," *U.S. Geol. Surv. Open-File Report 87-366* (1987): 5-7.

Klein, T.L., Day, W.C., "Tabular Summary of Lithologic Logs and Geologic Characteristics from Diamond Drill Holes in the Western International Falls and the Roseau 1° x 2° Quadrangles, Northern Minnesota," *U.S. Geol. Surv. Open-File Rept. 89-346* (1989): 17-19, 41.

Kreisman, Peter, and Poppe, Robert, "Introduction to Minnesota's Copper-Nickel Study with Aresource Overview," *Soc. of Mining Eng. of AIME Preprint No. 80-323* (1980): 9.

La Berge, Gene L., *Geology of the Lake Superior Region* (Geosciences Press, Inc., 1994), 313.

Lawler, T., "VLF-EM Geonics EM 16 Traverse, Little America Mine, Minnesota Dept. of Natural Resources, Hibbing, MN, Record

71-22-33, file item 1 (1983): 1.

Lawson, A.C., "Report on the Geology of the Rainy Lake Region," *Geol. Nat. Hist. Survey Canada Ann. Rept.* 3, pt. 1, Rept. F (1888): 183.

Lawson, A.C., "The Archean Geology of Rainy Lake, Restudied," *Geol. Surv. Canada, Mem.* 40 (1913): 115.

Lee, Carol Kindle and Kerr, S. Duff, "The Midcontinent Rift—A Frontier Province," *Oil and Gas Jour.* (June 12, 1984).

Lillie, R. J., *Parks and Plates: The Geology of Our National Parks, Monuments, and Seashores* (New York: W.W. Norton and Company, 2005).

Merriam, L.C., Jr., and Kurmis, Vilis (Eds.), *Voyageurs National Park, Minnesota: Research Studies of a New Park, Its Development, Potential Visitors, and Plant Communities, 1973-1980* (Univ. of Minn. College of Forestry, March 1981), 15.

Meuschke, J.L., Beooks, K.G., Henderson, J.R., Jr., and Schwartz, G.M., "Aeromagnetic and Geologic Map of Northeastern Koochiching County, Minnesota," U.S. Geol. Surv. Geophysical Investigations Map GP-133, 1957.

Minnesota Department of Natural Resources, *A Guide to Minnesota's Scientific and Natural Areas.*

Minnesota Geological Survey, "The Proposed Voyageurs National Park: Its Geology and Mineral Potential," *Minn. Geol. Surv.* (Minneapolis: Univ. of Minnesota, 1969), 16.

Minnesota Geological Survey, *Field Trip and Guidebook for the Western Vermilion District, Northeastern Minnesota* (Univ. of Minn. Guidebook Series No. 10, 1979).

Mooney, Harold M., Farnham, Paul R., Johnson, Stephen H., Volz, Gary, and Craddock, Campbell, "Seismic Studies Over the Midcontinent Gravity High in Minnesota and Northwestern Wisconsin," *Minnesota Geol. Survey Report of Investigations No. 11*, 1970.

Morey, G.B., Green, J. C., Ojakangas, R.W., and Sims, P.K., "Stratigraphy of the Lower Precambrian Rocks in the Vermilion District, Northeastern Minnesota," *Minn. Geol. Survey, Report of Investigations 14*, 1970.

Ojakangas, Richard W., "The Proposed Voyageurs National Park," *Minn. Geol. Surv.* (Minneapolis: Univ. of Minnesota, 1969).

Ojakangas, Richard W., "Rainy Lake Area," *in* Sims, P.K., and Morey, G.B., (Eds.), Early Precambrian, Chapter III, *Geology of Minnesota: A Centennial Volume* (Minneapolis: Minn. Geol. Surv., 1972), 163-171.

Ojakangas, Richard W., Meineke, David G., and Listerud, William H., "Geology, Sulfide Mineralization and Geochemistry of the Birchdale-Indus Area, Koochiching County, Northwestern Minnesota," *Minn. Geol. Survey, Report of Investigations 17*, 1977.

Ojakangas, Richard W., and Morey, G.B., "Keweenawan Sedimentary Rocks of the Lake Superior Region: A Summary," *Geol. Soc. Am. Memoir 156* (1982): 157-164.

Ojakangas, Richard W., and Morey, G.B., "Proterozoic Sedimentary Rocks [Abs.]," *Geol. Soc. Am. Memoir 156* (1982): 83-84.

Ojakangas, Richard W., and Olson, J.M., "Sedimentation and Petrology of Archean Feldspathic Quartzite and Conglomerate (Seine Series Equivalent), Rainy Lake, Minnesota," 28th Annual Inst. on Lake Superior Geology Proceedings (1982): 36-37.

Ontario Department of Mines-Geological Survey of Canada, Seine Bay Sheet, Rainy River District, Ontario, Ontario Dept. Mines-Geol. Surv. Canada, Aeromagnetic Series, Map 1150G, scale 1"= 1 mile, 1961a.

Ontario Department of Mines-Geological Survey of Canada, Fort Frances Sheet, Rainy River District, Ontario Dept. Mines-Geol. Surv. Canada, Aeromagnetic Series Map 1158G, scale 1"= 1 mile, 1962a.

Payne, Gregory A., "Water-Quality Reconnaissance of Lakes in Voyageurs National Park, Minnesota," *U.S. Geol. Surv. Open-File Rept. OF 79-556* (1979): 49.

Payne, Gregory A., "Water Quality Monitoring of Voyageurs National Park," *U.S. Geol. Surv. Prof. Paper 1275* (1981): 96.

Payne, Gregory A., "Water Quality of Lakes and Streams in Voyageurs National Park, Northern Minnesota, 1977-84," *U.S. Geol. Surv. Water Resources Investigations WRI 88-4016* (1991): 95.

Percival, J.A., Geology, Quetico, Ontario-United States, Geol. Surv. Canada Map 1682-A, 1988.

Percival, J.A., "A Regional Perspective of the Quetico Metasedimentary Belt, Superior Province, Canada," *Canadian Journal of Earth Sciences* 26 (1989): 677-693.

Percival, J.A., Stern, R.A., Digel, M.R., "Regional Geological Synthesis of Western Superior Province, Ontario, Part A," *Geol. Surv. of Canada, Paper 85-1A* (1985): 385-397.

Percival, J.A., and Sullivan, R.W., "Age Constraints On the Evolution of the Quetico Belt, Superior Province, Ontario," *in* de Witt, M.J., and Ashwal, L.D., *Workshop On Tectonic Evolution of Greenstone Belts, Planetary Institute Workshop LPI Technical Report 86-10*, 1986.

Percival, J.A., and Williams, H.R., "Late Archean Quetico Accretionary Complex, Superior Province, Canada," *Geology* 17 (1989): 23-25.

Perry, David E., "Gold Town to Ghost Town—Boom and Bust on Rainy Lake," Rainy Lake Interpretative Association (1993): 53.

Peterman, Z.E., "Petrology of the Metasediments of the Rainy Lake Region," M.S. Thesis, Univ. of Minnesota, Minneapolis, 1969.

Peterman, Z.E., and Day, W., "Early Proterozoic Activity on Archean Faults in the Western Superior Province—Evidence from Pseudotachylite," *Geology* 17 (1989): 1089-1092.

Poulsen, K.H., "The Geologic Setting of Mineralization in the Mine Centre-Fort Frances Area, District of Rainy River, Ontario," *Ontario Geol. Surv. Open File Rept. 5512* (1984): 129.

Poulsen, K.H., "Rainy Lake Wrench Zone: An Example of an Archean Subprovince Boundary in Northwestern Ontario (abs.)," *in Workshop on the Tectonic Evolution of Greenstone Belts, Lunar and Planetary Inst., Jan. 16-18, Houston Texas, Technical Rept. 86-10* (1986): 177-179.

Poulsen, K.H., Borradaile, G.J., and Kehlenbeck, M.M., "An Inverted Archean Succession at Rainy Lake, Ontario," *Canada Journal of Earth Sciences* 17 (1980): 1358-1369.

Schmidt, P.G., "Nature and Age of the Vermilion Fault System, Northern Minnesota," *Institute on Lake Superior Geology* 39, part 1 (1993): 69.

Schultz, K.J., "The Magmatic Evolution of the Vermilion Greenstone Belt, NE Minnesota," *Precambrian Research* 11 (1980): 215-245.

Schultz,-Ela, D., "Strain Patterns and Deformation History of the Vermilion District, Northeastern Minnesota," Ph.D. Thesis, Univ. of Minnesota, Minneapolis, 1988.

Sielaff, Richard O., Meyers, Cecil H., and Friest, Philip L., *The Economics of the Proposed Voyageurs National Park* (Duluth: Univ. of Minn., 1964).

Sims, Paul K., "Minnesota's Greenstone Belts—A New Exploration Target," *Skilling's Mining Review,* 58, no. 18 (1969): 1, 8, 9, 18, 19.

Sims, Paul K., Geologic Map of Minnesota, Minnesota Geol. Surv., Univ. of Minnesota, Minneapolis MN, Miscellaneous Map Series Map M-14, 1970.

Sims, Paul K., "Mineral Deposits in Lower Precambrian Rocks, Northern Minnesota," *in* Early Precambrian, Chapter III, *Geology of Minnesota: A Centennial Volume* (St. Paul: Univ. of Minn., Geol. Surv., 1972), 172-176.

Sims, Paul K., "Northern Minnesota, General Geologic Features," *in* Early Precambrian, Chapter III, *Geology of Minnesota*, 1972, 41-48.

Sims, Paul K., "Early Precambrian Tectonic-Igneous Evolution in the Vermilion District, Northeastern Minnesota," *Geol. Soc. Amer. Bull.* 87 (1976): 379-389.

Sims, Paul K., and Morey, G.B., "Minnesota Mineral Resources: A Brief Overview," *Skillings' Mining Review* 63, no.13 (March 30): 1-8.

Sims, Paul K., and Southwick, David L., 1985, Geologic map of Archean rocks, western Vermilion District, northern Minnesota, U.S. Geol. Surv. Misc. Investigations Series Map I-1527.

Sorensen, J.A., Glass, G.E., Schmidt, K.W., Huber, J.K., and Rapp, G.R., Jr., "Airborne Mercury Deposition and Watershed Characteristics in Relation to Mercury Concentrations in Water, Sediments, Plankton, and Fish of Eighty Northern Minnesota

Lakes," *Environmental Science Technology* 24, no. 11 (1990): 1716-1727.

Southwick, David L., "Vermilion Granite-Migmatite Massif," *in* Sims, P.K., and Morey, G.B. (eds.), *Geology of Minnesota: A Centennial Volume* (St. Paul: Minn. Geol. Surv., 1972), 108-119.

Southwick, David L., "On the Genesis of Archean Granite Through Two-Stage Melting of the Quetico Accretionary Prism at a Transpressional Plate Boundary," *Geol. Soc. Am. Bull.* 103 (1991): 1385-1394.

Southwick, David L., and Halls, H.C., "Compositional Characteristics of the Kenora-Kabetogama Dyke Swarm (Early Proterozoic), Minnesota and Ontario," *Canadian Jour. of Earth Sciences*, no. 24 (1987): 2197-2205.

Southwick, David L., and Ojakangas, Richard W., "Geologic Map of Minnesota, International Falls Sheet, Minnesota Geol. Surv., scale 1:250,000, 1979.

Southwick, David L., and Sims, Paul K., 1980, The Vermilion granite complex—a new name for old rocks in northern Minnesota, *in* Shorter Contributions to Mineralogy and Petrology 1979, U.S. Geol. Surv. Prof. Paper 1124-A, A1-A11.

Southwick, David L., and Day, W.C., "Geology and Petrology of Proterozoic Mafic Dikes, North-Central Minnesota and Western Ontario," *Can. J. Earth Sci.* 20, no. 4 (1983): 622-638, doi: 10.1139/e83-058, © 1983 NRC Canada.

Sullivan, D.W., "Private Report on Grassy Portage Bay Property to Seemar Mines Limited, March 8, 1970, unpub. rept., Resident Geologist's files, Ontario Ministry of Natural Resources, Kenora, Ontario.

Tabor, J.R., "Deformational and Metamorphic History of Archean Rocks in the Rainy Lake District, Northern Minnesota," Ph.D. Thesis, Univ. of Minnesota, Minneapolis, 1988.

Tabor, J.R., and Huddleston, P.J., "Deformation at an Archean Subprovince Boundary, Northern Minnesota," *Can. Jour. Earth Sci.* 28 (1991): 292-307.

United States Geological Survey, Topographic Map of Voyageurs National Park, Minnesota, U.S. Geol. Surv. Special Topographic Map, 1:50,000 (1 sheet), 1979.

White, Walter S., "Geologic Evidence for Crustal Structure in the Western Lake Superior Basin," *in The Earth Beneath the Continents* (1966): 28-41.

Williams, H.R., "Quetico Subprovince," *in* Thurston, P.C., Williams, H.R., Sutcliffe, R.H., and Stott, G.M., eds, *Geology of Ontario, Ontario Geol. Surv. Special Volume 4, Part 1* (1991), 383-403.

Williams, H.R., Stott, G.M., Heather, K.B., Muir, T.L., and Sage, R.P., "Wawa Subprovince," *in* Thurston, P.C., Williams, H.R., Sutcliffe, R.H., and Stott, G.M., eds, *Geology of Ontario, Ontario Geol. Surv. Special Volume 4, Part 1* (1991), 485-539.

Winchell, H.V., and Grant, U.S., "Preliminary Report on the Rainy Lake Gold Region," *23rd Annual Report, State of Minnesota* (1895 and 1899): 36-105.

Wood, J., "Epiclastic Sedimentation and Stratigraphy in the North Spirit Lake and Rainy Lake Areas: A Comparison," *Precambrian Research*, 12 (1980): 227-255.

Zoltai, S.C., Surficial Geology, Kenora-Rainy River, Ontario Dept. Lands and Forests, Map S165, scale 1"=8 miles, 1965.

INDEX